Stylianos Tsaparas

Spatially Transformed Social Networks Model

AF140965

Stylianos Tsaparas

Spatially Transformed Social Networks Model

An entirely new Urban Planning approach

LAP LAMBERT Academic Publishing

Impressum / Imprint
Bibliografische Information der Deutschen Nationalbibliothek: Die Deutsche
Nationalbibliothek verzeichnet diese Publikation in der Deutschen
Nationalbibliografie; detaillierte bibliografische Daten sind im Internet über
http://dnb.d-nb.de abrufbar.
Alle in diesem Buch genannten Marken und Produktnamen unterliegen
warenzeichen-, marken- oder patentrechtlichem Schutz bzw. sind
Warenzeichen oder eingetragene Warenzeichen der jeweiligen Inhaber. Die
Wiedergabe von Marken, Produktnamen, Gebrauchsnamen, Handelsnamen,
Warenbezeichnungen u.s.w. in diesem Werk berechtigt auch ohne besondere
Kennzeichnung nicht zu der Annahme, dass solche Namen im Sinne der
Warenzeichen- und Markenschutzgesetzgebung als frei zu betrachten wären
und daher von jedermann benutzt werden dürften.

Bibliographic information published by the Deutsche Nationalbibliothek: The
Deutsche Nationalbibliothek lists this publication in the Deutsche
Nationalbibliografie; detailed bibliographic data are available in the Internet
at http://dnb.d-nb.de.
Any brand names and product names mentioned in this book are subject to
trademark, brand or patent protection and are trademarks or registered
trademarks of their respective holders. The use of brand names, product
names, common names, trade names, product descriptions etc. even without
a particular marking in this work is in no way to be construed to mean that
such names may be regarded as unrestricted in respect of trademark and
brand protection legislation and could thus be used by anyone.

Coverbild / Cover image: www.ingimage.com

Verlag / Publisher:
LAP LAMBERT Academic Publishing
ist ein Imprint der / is a trademark of
OmniScriptum GmbH & Co. KG
Bahnhofstraße 28, 66111 Saarbrücken, Deutschland / Germany
Email: info@lap-publishing.com

Herstellung: siehe letzte Seite /
Printed at: see last page
ISBN: 978-3-659-82235-3

Zugl. / Approved by: London, UCL, Diss., 2014

Acknowledgements

My warmest thanks to Dr Martin Zaltz Austwick, whose patience, suggestions and directions led me to fascinating and unknown academic readings and understandings that turned this study into my own space of learning. I would also like to thank Kinda Al-Sayed for helping me access the space syntax literature in a deep and meaningful way. Finally, I would like to thank from the bottom of my heart my own social network, and especially my parents, brothers and my partner Valia because none of this would be possible if it wasn't for them. I thank them for their openness and willingness throughout my dreams.

Table of Contents

1. Introduction

In the beginning of the 21st century, crisis is an enormous challenge for the sustainability of cities. The impact this new global situation has in our daily urban life could be more understandable if we dove into the nature of crisis.

Very shortly, in terms of historical evolution, transitions from one era to another, have most of the times caused conditions of crisis because the world's interpretation models were in question. Technological and scientific progress played catalytic role to this questioning. For example, there would not have been the Enlightenment of the 18th century, if it had not been for the revolutionary scientific discoveries of Copernicus and Galileo during the Renaissance of the 16th century. And there would not have been Marxism in the second half of the 19th century, if the Industrial and Technological Revolution had not been preceded earlier. Of course, transitions from one era to another, as I mentioned earlier, have caused crisis and revolutions.

In other words, if on one hand Enlightenment introduces the individualistic model of "Human" as the model of interpretation of the world (and consequently of urbanism) and on the other hand, Marxism introduces the holistic model of "Community", nowadays in digital revolution, *what formula can be a satisfactory model for interpreting and confronting the problems of contemporary production of urban space?*

In a sense, digital revolution and its products introduces a new interpretative model of the world, based on the tremendous power of social networks (Pentland, 2014). Such is the power of our technology that we can analyse, visualise and refer to communities in terms of individual connectivity. It is a great challenge for humanity to discover the new model's operational rules and to convert them into scientific tools in order to shape the contemporary city in a sustainable way.

Under this point of view, some questions can be erected:

What type/architecture of citizens' connectivity leads to an urban sustainable system?

Could the answer in that problem be a base for a solution of a plethora of urban problems?

There is an Agency-Structure problem here: How does a social structure emerge?

1.2. The vision and the area of research

Considering that urban space is a significant social product, the research interests are focused on re-reading the approach of the urban design process by utilising the theory of Social Networks (SN). The research proposal inserts the social network as a vivid mutable and independent organisation whose shape and architecture is in the powerful position to influence the structure and the function of a city.

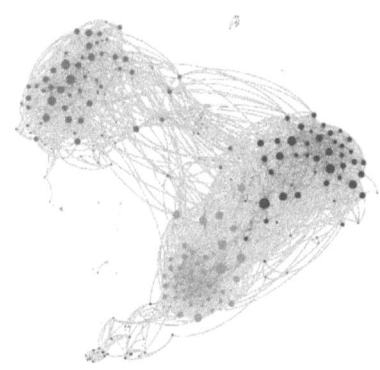

If there is a correlation between the position of a node in a social graph and its spatial position then the handling of some nodes' connectivity in an appropriate social network could become a creative solution in policy and thus, urban design problems in the context of the contemporary smart city concept. The research project tries to decode this particular relation and to enhance the conviction that the qualitative upgrade of urban space can possibly occur through strategic interventions in the architecture of appropriate social networks. This is in contrast to the dominant trend of direct interventions in the real urban body.

Aligned with that hypothesis is the need for quantitatively oriented research methodology in order for the impact of the investigation to be increased. Complexity theory and its techniques, such as Agent Based Models (ABM) in which simple agents, by following simple rules, could generate amazingly complex structures could make the new interpretative model feasible.

Finally, it is crucial to distinguish the difference between the term Social Networks and the term Social Media. In the first case, the phrase refers to the real world relationships among people or groups which are depicted with graphs, with nodes or agents for the individuals or groups and lines for the relationships. On the other hand, social media, like Facebook or Twitter, are platforms that host cyber connections among internet users. This particular study focuses on Social Networks.

1.3 Research method in brief

This book introduces the Spatially Transformed Social Networks (STSN) model which is an abstract Agent Based Model (ABM) designed for the needs of the research. The aim is to simulate space production processes by generating connected agents in social networks. The input is a dataset containing coded information regarding different types of social networks, while the output is abstract spatial forms.

The model essentially embeds two crucial rules which are followed by the agents:

1. Three degrees of influence.

This rule indicates the social distance between two individuals and means that each one could influence only a friend, a friend of that friend and a friend of the friend of that friend. The percentage of the impact is reduced when the social degree of separation is increased.

2. Homophily.

It describes the social phenomenon according to which, if two people have similar characteristics then they are more likely to be connected and vice versa. It is caused by a phenomenon called structural location, according to which people with similar interests are more possible to be at similar locations and again vice versa.

"Homophily" moves the agents and the "Three degrees of influence" is the criterion for the creation of spatial connections among them.

The combination of these two simple rules produces a complex system of spatially distributed objects. The next step is the analysis of this spatial structure and its comparison with the social network's structure from which is generated.

1.4 Structure of the report

At the beginning of this report is the introduction which helps the reader to understand my deepest anxieties and the background of the my research interests. The research definition is the next chapter and the third is the literature review. There, I present some key concepts from the wider area of sociology as well as some information related to the social network analysis field. Additionally, there is a literature review of the field of space syntax and generally from the complexity theories of cities. In the methodology chapter I present in detail the process I followed as well as I explain the structure of the STSN model. In the results, I analyse, compare and interpret the outputs of the whole process. The conclusions of this procedure are presented in the final chapter of the report, together with the potential implications, the limitations and the further research.

2. Research Definition

This chapter presents the context in which the research interest is constructed. The target here is to make the motivation, the aim and the objectives of this study as clear as possible.

2.1 Motivation

The process of urban design in the majority of cases is from an "airplane perspective" and thus, almost based on intuitive criteria applied by authority centres and individual experts, like architectural firms and urban designers in a top-down logic. Thus, these centres draw lines on white papers hoping that the sense they have for reality and the needs of citizens is right. In fact, any urban design proposal, like any model, is only as good as its assumptions and unfortunately, in these cases there are not enough evidences that the proposed solutions articulate them efficiently.

On the other hand, millions of words have been written about cities as complex systems during the last decade, which introduce the concept of emergence of urban patterns (Portugali et al., 2012b). Contemporary theories consider cities as dynamic systems under an unstable equilibrium and evolution processes and therefore they grow in a bottom-up logic (Batty, 2005). Consequently, scientific interests are focused on how relationships between parts of systems shape collective behaviours and at the same time how these systems form relationships with their environments (Batty, 2008).

However, in the way the scientific community understands cities today, all this progress is not yet related with a constant architectural or urban design practice oriented to compose high quality and mainly sustainable urban spaces. In other words, for the time being, a lot of interest is spent in the analysis of urban pattern emergence and less in how to use it for the needs of architectural and urban design. Therefore and diving deeper in this particular research, as it will be discussed in the literature review, there is no rigorous novel focused on how social networks topology could shape spatial structures, given the fact that citizens are vital parts of every city.

2.2 Aims

The aim of this research is to investigate potential correlations between specific types of social networks and emergent spatial patterns.

Specifically, the research question could be:

To what extend the way that social ties are articulated could be an explanatory variable for the formation of space?

2.3 Objectives

The objectives of this particular study is to develop an Agent Based Model (ABM) which by using data could emerge spatial patterns. These patterns should be analysed and compared with social measurements. The output should answer to the initial question.

3. Literature Review

3.1 Why Social Networks (SN)

The process of urban design and growth is a tremendously complicated procedure as it encapsulates a plethora of considerable aspects: from commercial uses to residential needs and from health problems to aesthetic issues. The extent to which every aspect derives from and addresses human needs, social networks should be a common platform for all these parameters. By researching and modelling SNs it is possible for the built environment scientists to generate models that can explain, simulate and predict urban phenomena in terms of cities' growth.

After examining sets of data from many scientific fields, Christakis and Fowler (2009) exported some very interesting findings about the qualitative principles from which the nature of SNs derived. These fundamental characteristics are crucial in understanding what the role of SNs could be in the timeless process of urban development and in starting to think the city as a potential product of them. Similarities between these two "organisations" –cities and SNs- are worth noting and are presenting below.

First and foremost, every network, like very city, is considered as a superstructure, a vivid organisation, with its own anatomy and physiology which is independent of its members. SNs can have properties and functions -they have memory of their structure and can be auto-copied over space and time- which cannot be controlled or even perceived by its members. As an idea, it is interesting that the life of SNs lasts more than the life of their members and so they have their own life, they grow, they change, they survive and die. Social networks can often be "re-self-stretched". They also have the ability to fill their blanks in the same way that a wound is healed (Christakis & Fowler, 2013).

Various things flow and move in them. They can introduce a kind of intelligence that magnifies or completes individual intelligence. Social networks can receive and contain information such as reliability rules and do calculations that congregate millions of decisions such as the choice of the most advantageous urban intervention. Finally, networks can have these outcomes regardless of their members' intelligence (Coleman, 2007).

3.1.2 Structure matters

Diving deeper in SNs theory, and investigating more the Christakis and Fowler (2009) work, an impressive characteristic is that their topology -or architecture- is directly related to their properties and function. In a sense, the position an individual holds in the graph -as a node- is a determinant factor which regulates the impact these properties have in the individual. Consequently, the way each one of us interacts with others or with our environment, is more related to who we are connected to rather than to who we are. Christakis and Fowler are going further with their arguments by introducing two proposals as paradigms of what this property could mean. They focus on the criminality of a city and suggest that in order for it to be reduced as a negative urban phenomenon, authorities should intervene in the types of connections among potential criminals and optimise them. Furthermore, for poverty to be reduced, it is necessary to help the poor to establish new relationships with other members of society. By aiming at the periphery of a network in order to re-link its members, the benefits are both for the vulnerable members and the whole network.

Under this point of view, a moral issue is erected: what are the limits of our intervention in humans' lives since peoples' relationships are in the core of the proposals? Additionally, if our actions are affected by the properties of the networks we belong to, how free are we? However, this is not the subject of this study.

These properties can be comprehended only by the study of the whole group and not by the individual study of the persons. In a sense, the study of the impact that SNs could have is a fantastic challenge, the answer of which could lead to a powerful tool for the function of the contemporary City. This is a fundamental way to understand the urban space production: it is not about a decision of an architect or an urban planner. Probably, we should take into account appropriate networks as vivid organisations which shape the city (Smith, 2012). Consequently, the understanding of the structure and operation of SNs, as well as the identification of collective properties that could not be identified in their members separately, should be the axis of a new research approach in the sector of urban planning and design. Undoubtedly, this approach introduces the emergence of patterns in a bottom-up process and thus belongs to the general scientific field of the Complexity Theory of Cities (CTC) (Portugali, 2012a) that will be examined in a later section of this chapter.

The following chapter discusses existing literature related to the human behaviour from the perspective of the scientific field of Sociology. The focus of the bibliographic research is on finding the specific knowledge that brings to light the organisational principles of social networks. The following section examines the fundamental rules that characterise the latter in an attempt to clarify the way that they work and mainly how the members of a network interact with each other. The aim is to adopt these rules in order to build and enhance the reliability of the model that this research introduces.

3.2 Social Networks Analysis (SNA)

During the past decades, SNA has become a popular method for approaching qualitative and quantitative concepts in social scientific work. However, it is impossible for a whole scientific field such as graph theory or sociology to be quoted in a few lines. In what follows, some of the most fundamental aspects of the SNA field are selectively presented.

3.2.1 Six degrees of separation and three degrees of influence

Half a century ago, Stanley Milgram and Travers started looking for a measurement which could indicate how connected people are (Milgram & Travers, 1969). According to a famous experiment that was held in Nebraska, USA, they argued that each one of us is connected with all the other people in a means of six degrees depth or "six degrees of separation". This impressing phenomenon is also known as "small world" (Kochen, 1989). In the core of this finding was the idea that between two individuals could be approximately five people.

However, a portion of the academic world expressed its scepticism on whether the results of the above experiment had global power or were only reliable in a local scale. This contradiction ended in 2002 when Peter Dodds and Roby Muhamad repeated Milgram's experiment in a global scale via the new technology of electronic mails (Dodds et al, 2003). Given the fact that in this particular global experiment 98,000 people involved, the new results were impressive as the two researchers found that the rule of the six degrees separation did not change.

On the other hand, even though the world seems small indeed and well connected, the number of people that an individual can influence is much lower. Recent studies show that the spread of the impact in a SN is following the rule of the "three degrees of influence" (Kenrick, Goldstein & Braver, 2012). This rule indicates the social distance between two individuals and means that everyone could influence only a friend, a friend of that friend and a friend of the friend of that friend. More specifically, studies in loneliness dissemination in whole SNs show that the probability of an individual to be influenced by another with whom he/she is connected in 1st degree, is 52%. The probability for connections of 2nd degree is 25% and for 3rd degree relationships is 15% (Christakis, 2009).

According to Nicholas Christakis and James Fowler, there are three potential reasons for the elimination of the impact on the 4th degree and on. Very briefly, in a social level the accuracy of the information which is spread through a SN is descending. This phenomenon is called inherent attenuation. Secondly, the impact is dramatically reduced due to the evolution of SNs which makes the links beyond the 3rd degree unstable. This explanation is known as network unsteadiness. The last reason for the limitation of influence is known as evolutionary aim which seems to be due to the fact that for thousand of years, human beings were organised in small communities in which there were no links beyond the 4th degree.

In conclusion, on one hand the six degrees of separation between two randomly selected individuals is an indicator of the level of connectivity of a SN and on the other, the three degrees of influence reflect how contagious a SN is. The last rule is a significant part of the way human SNs work which raises complexity and constraints in people's connectivity despite the progress of technology and social media. These two properties characterise the structure and the function of every SN and are considered as the anatomy and the physiology of the human super-organism.

3.2.2 Basic principals of social connections

In what follows, an epigrammatically short mention of some basic principals regarding social networks' connections is presented. The most related aspects to this particular study are the phenomena of Homophily and Propinquity which will be examined in more detail in a later subsection.

To begin with, networks are a set of relationships which are depicted in "sociograms" (Moreno, 1953) or what mathematicians call, graphs. The latter consist of a set of objects -in mathematical terms, nodes- and a mapping of relations between them expressed by lines. The simplest network is a dyad which could express undirected, mutual or directed connections. Shortly, a link is called undirected when there is no need to add arrows to the symbol of tie. Usually, it reflects a coexistence of two individuals in the same room. Mutual is a connection in which a flow of something between two nodes has equivalent weight but opposite directions and thus reflects symmetric relationships. Directed connections are those in which there is flow of something from one node to another but not the opposite (Kadushin, 2011).

Finally, depending on the number of relationships between a pair, such as partners and friends, the link is called multiplex (Verbrugge, Lois M., 1979).

3.2.2.1 Homophily

Homophily comes from the Greek ομοφυλία, "love of the same" and is mentioned in bibliography as birds-of-a-feather-flock-together. The concept was introduced into social theory by Lazarsfeld and Merton (1978) and describes the social phenomenon according to which if two people have similar characteristics then they are more likely to be connected and vice versa (Verbrugge, 1977).

The process of homophily is a chicken-and-egg situation which is caused basically by two factors. The first cause is that common contacts could lead to common behaviours and common behaviours may link people with common attributes (McPherson, Smith-Lovin and Cook, 2001). The second and more interesting factor is structural location which is when people with similar interests are more likely to be at similar locations and again vice versa (Feld and Carter, 1998). Indeed, it is more likely for someone to find people interested in architecture in a school of urban design than in a medicine class.

Kadushin expresses the sum of the homophily's attributes in a simplified and clear sentence:

> *"(...) if people flock together, it appears that there are four processes involved: (1) the same kinds of people come together; (2) people influence one another and in the process become alike; (3) people can end up in the same place; (4) and once they are in the same place, the very place influences them to become alike" (Kadushin, p. 20, 2011).*

In other words, social structures and spatial locations are correlated. This is a very important finding exported by literature in the field of sociology which along with the propinquity principle presented in the next section, is the backbone of this particular research.

3.2.2.2 Propinquity

Further underlying the importance of location in SNA, a mention to the social phenomenon called propinquity – which comes from the Latin propinquitas, "nearness"- is necessary. According to this, people are more likely to be connected if they are geographically close (Adams & Allan, 1999). Propinquity or proximity could have and reflect both physical and psychological aspects which lead to interpersonal attraction. Indeed, Festinger, Schacter and Back (1950) demonstrated that in many cases friendly pairs were generated because of the spatial proximity between the nodes. The differences between propinquity and homophily are thin and in many cases homophily is considered as a different kind of propinquity.

3.2.3. Key metrics of networks or "Distributions"

SNA follows the fundamental principles of Graph theory which is an interesting branch of mathematics and a powerful tool for further understanding and manipulating large and complex sociograms (Harary, Norman and Cartwright, 1965). What follows is a presentation of some key metrics that Graph theory offers, related to network structures, the use of which contributes to the quantification of the characteristics of the initial social networks.

A basic concept in graph theory, is the degree (or valency) of a node. In a network the degree of a node is the number of its neighbours with which it is connected (Diestel, 2005). It is a structural concept and gives shape to every network.

A metric related to the degree is the connectivity. Connectivity is an important measure of how strong a network is. It refers to the minimum number of edges or nodes that are essential in order for the network to stay coherent (Diestel, 2005). In simpler words, if these nodes or edges be removed, the network will be disconnected.

Another significant concept of SNA is centrality. In simple words, centrality is related to nodes with the minimum path lengths from all other nodes. In a sense and in real life terms, centrality could refer to the most influential person(s) in a network. The agents which are involved in many ties are called central agents. This characteristic increases the visibility of these nodes from other agents in the network. As it is expected, the centrality of an agent increases when it becomes the object of more ties. However, this is not necessary in directional relations which are the cases when an agent initiates the ties by itself.

Furthermore, in undirected graphs -the kind of networks this study focuses on- a path is defined as the finite or infinite sequence of edges that connect a sequence of nodes which are distinct from one another (Korte et al, 1990). The shortest path between two nodes in a network is called distance.

Finally, graph density is an indicator that reflects how close to the maximal number of edges is the actual number of edges. In other words, it is the proportion of links in a network in relation to the total number of links that is possible (Xu, Guandong et al, 2010).

3.3 Complexity Theories of Cities (CTC)

In the core of interest of this study is the complexity theory of cities (CTC) (Portugali, 2011). During the second half of the 20th century the dominant trend of how the city looks like was that a city is a machine, that is it consists of numerous of pieces which are articulated in harmony and the total system is always in equilibrium (Marshall, 2009). Today, the analogy of the machine is based on evolving biologies or ecosystems because cities develop occupying the available space like biological organisms (Batty, 2010).

One of the most popular and well defined theories related to CTC, to the extent that it attempts to answer to the general question of how structures emerge, is the generic city of Bill Hillier (Hillier, 2012). His life-long studies on the spatial and functional structures of cities refer to the Space Syntax which is a mathematical patchwork useful for the designers and planners of all scales of planning. It contributes to the understanding and testing of the links between space and society by investigating the laws under which the relation is shaped.

More specifically, Hillier and his partners attempt to extract knowledge regarding the connection and reaction of the two constant systems that every city consists of: the physical system -buildings, roads, infrastructure- and the human system -movements, interactions, activities. In the context of this relationship, the core of space syntax's philosophy is based on networks and their theory. Hillier argues that spatial configurations can be explained and analysed by the study of spatial and functional networks which constitute a fundamental aspect of any city. In that sense, he proposes a new definition of a city "as a network of linked centres at all scales set into a background network of residential space" (Hillier, 2012 p. 2).

An interesting result that his research produced and is crucial for this particular study is that the differences between micro-economic and socio-cultural forces, which coexist in a city, are subject to the geometry and scale of its street networks. These two variables are dependent of each other.

In that sense, and by using the least line maps of cities (Hillier and Hanson, 1984), a technique developed by space syntax, Hillier correlates the length of lines with the angle they are connected at their ends:

> *"the longer the line, the more likely it is to end in a nearly straight connection to another line" while "the shorter the line, the more likely it is to end in a right angle or near right angle" (Hillier, 2012, p. 2).*

3.4 Related Work

It is hard for someone to find relevant research to focus on how social networks are transformed into spatial patterns. Wong et al. (2005) generated an exponential random model to study random graphs. The scenario they use to perform a simulation was random scattered vertices and they study the effect of the neighbourhood radius on small-world structures. However, they did not do any statement related to hierarchical networks or Barabasi-Albert models.

4. Methodology

This chapter presents the methodology the research followed. The target is to provide critical information about a framework of significant steps-decisions that were taken into consideration. A clearer picture of this procedure will facilitate the understating of the results as well as of the model's limitations.

In short, the starting point of the research is the double observation that cities as complex systems consist of complicated relations and the secret power of social networks. These points are developed in more detail in the section of the Research Definition.

The next step is the investigation of related scientific literature, the main findings of which are mentioned in the Literature Review chapter. Briefly, the research focused on two general directions: the bibliography which refers to the field of sociology and social networks and the field of pattern emergence in urban spaces.

This chapter in turn contains information about crucial parts of the model's operational rules as well as of the SNA and the spatial analysis of the model's outputs. Diagrammatically, the followed process is divided in five main axes:
1. Specific types of social networks are generated in Gephi software, analysed and exported as quantified data.

2. The data are imported in the STSN model and the model in turn exports a spatial graph by applying specific rules to all agents.

3. The spatial graph is imported to the UCL DepthMapX which runs Visibility Graphs Analyses (VGA).

4. The quantitative measurements of the VGA are compared with the statistics of the SNA which were exported from the first stage.

5. A calibration of the model occurs through the multiple iteration of these four steps.

Calibration allows the export of relatively safe final conclusions regarding the relationships between social networks' structures and their spatial expressions in terms of VGA as well as it encourages the comparative study between different spatial structures.

Last but not least, the results trigger a discussion regarding how expected the findings are or not, an interpretation of them and potential implications. Simultaneously, this section identifies limitations of the model and suggests further research.

What follows is a detailed presentation of the five steps mentioned above, the depth of which depends on the length constraints of the study.

4.1 Building the social networks

Like every model that attempts to interpret a part of reality embeds a series of decisions and assumptions, so is the STSN model aligned with that. Below, there is a mention and description of these.

4.1.1 Assumptions

Assumptions usually define the context of the research by chiseling the field of interest as well as simplifying the parameters that the model should take into account. In this particular model, the assumptions are described below.

I. Structural concept.

In the core of this research there is an animated interest in linking social networks with their spatial imprinting. Quantified metrics directly correlated with the geometry of these structures constitute the objectives of this particular study. Hence, cultural concepts regarding the what and why of the emergence of social networks are not a subject of this research.

II. Unweighted connections.

The model considers the quality of all ties as equal for all nodes. Thus, the model is simplified.

III. No gender

Sociologists tend to consider that gender matters in the process of information diffusion (Kirke, 2009). Males and females tend to trust more individuals of the same gender. A no-gender decision eliminates the possibility of selective behaviour based on cultural characteristics.

IV. Same level of motivation.

Another cultural aspect regarding the emergence and operation of social networks which is not a subject of this research is the level of motivation for individuals to create new connections. All agents should be equally motivated and act with a common denominator in order for results to become clearer.

V. Not multiplex relationships.

In every day's routine, each individual is invited to play multiple social roles, from the role of son or parent to the role of professional partner or citizen. Thus, the same individual could belong to multiple social networks, usually of different structures, each of which is partially overlapping in many cases. This phenomenon is called multiplexity by sociologists (Wardhaugh, 2006). In a sense, multiplex ties can enhance a relationship or create conflicts of interests and thus this behaviour is in contrast to the unweighted connections. Furthermore, although multiplexity is an extremely interesting field, there is limited formal theoretical background in order for a quantitatively hypothesis on that field to be tested.

VI. Reciprocal ties.

A very important assumption that the STSN model adopts is the undirected connections. Every agent should be in the position to send or receive information from all the other agents with which it is linked.

VII. Typically symmetric relationships.

By this phrase sociologists describe a relationship in which all members have equal level of authority over the others. Such kind of relationships are friends, neighbours and coworkers. Typically asymmetric connections is the opposite situation and is hard to analyse, except for the fact that it implies weighted and directed connections.

4.1.2 What social networks?

For the needs of this study, three are the types of social networks that are recorded and examined in relation to their spatial transformation: The Erdős–Rényi (ER) model, the Hierarchical Network (HN) model and finally the Barabási–Albert (BA) model. The selection of only these types does not exhaust the range of graph's theory bibliography, however it can be representative. Fundamental characteristics and crucial differences among them are the criteria for adopting these kind of structures and are presented below:

I. Erdős–Rényi (ER) model

The ER model belongs to the wider category of random graphs. According to graph theory, the edges of networks are added in a random way and thus every node has the same and independent probability to be a pair with another (Newman et al., 2001). Additionally, random graphs generally do not exhibit power law degree distribution.

Although the ER model is abstract and thus not realistic enough, the lack of constraints makes the ER model to be considered as a basic conceptual platform for the needs of this research.

II. Hierarchical Network (HN) model

The fundamental characteristic of HN models is that they grow by replicating an initial cluster of the network. In contrast to ER models, HNs are scale-free networks and appear to have proportionally more nodes in the role of a hub. However, the main property of that type of networks is that they combine scale-free topology and high clustering into one single model (Ravasz & Barabasi, 2003). These principles of HN models, in addition to the fact that several real-life networks belong to this category - from World Wide Web and armies to some human social networks- are the reason for including HNs in this study.

III. Barabási–Albert (BA) model

The BA model is the third type of networks that is examined in this book. The crucial difference from the ER and HN models is that not even it presents a power law degree distribution but also it incorporates the important concept of growth and preferential attachment (Albert et al., 2002). Growth means that instead of generating a set of nodes from the beginning, the algorithm increases the number of nodes in the network over time. The preferential attachment is a probability rule which means that the more connected a vertex of a graph is, the more likely it is to receive new ties. As a consequence, vertices with high number of edges have a greater probability to attract a new link when it is added to the network. In a sense, it is a kind of gravity.

Both these phenomena exist widely in real networks. For instance, in terms of real human social networks, heavily connected people reflect well-known personalities with lots of relations. It is observed that a newcomer who enters the community intuitively tends to be linked to one of those more popular people rather than to an unknown. The same principle can interpret the phenomenon of "the rich get richer" as well as it could, in a sense, explain social behaviours accompanied by spatial extensions such as homophily and propinquity.

Furthermore, the BA model, in contrast to the HNs in which the size of the network is independent of its average clustering coefficient, predicts that the higher the number of nodes, the lower the clustering co-efficient is expected to be presented (Dorogovtsev et al., 2002).

4.1.3 Quantifying the topology of the network

In the literature review chapter there is a mention of some basic concepts of graph theory regarding the quantitative statistics of networks' topologies. The metrics that are useful and convenient to the needs of this research are presented below. See diagrams in Appendix C.

I. Average degree and degree distribution.

Every node is connected to another and the total amount of its neighbours in undirected graphs is called degree. The average degree is the average of these connections (Albert, 2006). In ER models, average degree is considered as an indicator for how weak or strong social ties among individuals are. The higher the average degree is, the more connected a social group is.

The degree distribution is a useful measure of how many nodes have a specific number of edges. Therefore, it is a tool for investigating whether a structure is scale-free or not.

II. Network diameter.

The diameter measures the distance between the two most distant nodes in the graph. Connected nodes have graph distance 1 (Albert, 2006). Distance is called the average graph-distance among all links of vertices and is the base of two measures. the Betweenness Centrality and the Closeness Centrality.

III. Betweenness centrality.

This measure indicates how often a vertex appears in shortest paths among agents in the network. It seems to be very useful in identifying well connected vertices.

IV. Closeness centrality.

The target of this statistic is to find the average distance from a given starting vertex to all other vertices in the network (Albert, 2006).

V. Eccentricity.

Eccentricity refers to the distance from a given starting vertex to the fastest vertex in the network (Albert, 2006).

VI. Graph density.

Measures how close the network is for it to be completed. A complete graph has all possible edges and density equal to 1 (Albert, 2006).

VII.Modularity.

Modularity is a very interesting measure because it detects the number of local communities in a network. The number of communities is directly correlated with the resolution (Lambiotte et al, 2008) that is the least limit of edges per node which classifies the size of communities.

4.1.4 Data Preparation

In this stage, two datasets are generated by using Gephi software, 0.8.2.beta (Bastian et al., 2009). Gephi is designed for academic use and its main purpose is to build, analyse and visualise social networks. For the needs of this book, these two datasets concern the creation of a HN (Balanced Tree) and a BA model, both articulated by 512 nodes. This number is determined by the degree of the root of the Balanced Tree network, which is 2, in combination with the height of the tree which is 8. Further information about quantitative characteristics of the networks is hosted in a later section.

Gephi exports the datasets in a Comma Separated Values (CSV) format which contains the ID of the nodes. Every row starts with the ID of a node with which the other nodes in the same row are connected. Thus, if the first number of a row r is A, the second number of r is B, the third is C and the fourth is D, the sequence of the pairs are A-B, A-C, A-D.

A, B, C and D are separated by semi columns. In order for the STSN model to be able to read the dataset, these values should each belong in a unique column. In other words, manual modification of the dataset in Microsoft Ecxel or a similar software is required. The final output should be a CSV file again with A, B, C and D values belonging to four discrete columns.

These modified datasets will be the input of the second version of the STSN model the function of which is described in the next section.

4.2 Building the STSN model

4.2.1 Cellular Automata (CA) versus Complex Agents

Cities are considered nowadays as complex systems. This proposal derives from the fact that urban objects such as land value, transportations etc. are determined by the way that each of them interact with its nearest neighbours. On the other hand, every network is defined by its members' interactions. Since the cell's relation to its nearest neighbour is in the core of the CA's concept, which was first introduced by Von Neumann (1951), the choice of using a CA model seems attractive. In a CA model all the cells are updated simultaneously according to what is the value of their neighbours and a set of rules. This property indicates the ability of CA to easily combine changes in space, time and state and thus to be considered as ideal models for large - scale computer simulations (Zheng et al, 2009). However, this kind of models pose a serious constraint for the needs of this particular study. The power that performs the STSN model is the SN's structure and thus the relation and interaction among agents. As it will become clearer in a later section of this chapter, this interaction affects not only the neighbours but also the neighbours of the neighbours and the neighbours of the neighbours of the neighbours. Although CAs can simulate indirect relations are not a preferably methodological approach because CA has some restrictions in their cell shapes which reduce their ability to simulate complex environments. For instance, each cell is geometrically parallel with all other cells. In that way cells can form only orthogonal spatial relations.

Agent Based Models (ABM) come to bridge this limitation by emphasising to interactions among individuals which could shape non-orthogonal and complex environments. ABMs simulate the behaviour of the latter (agents) which derives from a set of rules that govern their relation with other agents as well as with their environment. These rules are usually formed as a sequence of decisions -decision tree (Torrens, 2012). The most interesting thing in these types of models is that, by following relatively simple rules, autonomous agents produce incredibly complex outputs, characterised by emergent patterns. In a sense, these individual agents by following a route of decision making processes "develop" a kind of intelligence which in turn produces a heterogeneous behaviour.

One negative consequence of this process which is considered as a limitation, is that in many cases, ABMs require high computational power in order for the model to be performed (Bonabeau, 2002). Another issue arising from the literature of CTC is that there is no rigorous definition for the nature and the meaning of the urban agent although Batty (2005) attempts to specify one. According to him agents are:

> "...objects that do not have fixed location but act and interact with one another as well as the environment in which they exist, according to some purpose. In this sense agents are usually considered as acting autonomously..." (Batty ibid., pp. 209).

> However, he continues by mentioning that "...autonomous agents thus
> cover a wide variety of behaving objects from humans and other animals
> or plants to mobile robots...".

Franklin and Graesser (1997) in their very interesting paper Is it an agent or just a program? A taxonomy for autonomous agents provide two general classifications regarding the agents' nature and the way they act and sense. According to this work, there are the passive agents that simply interact with their environment and the cognitive agents that not only interact with what they encounter but act by using some protocols and goals. Although it is not clear where the line between these two categories is in practice (Portugali, 2012a), the STSN model seems to be based on the second group because of the main concept behind the model which is the fact that the agents belong to a social network and thus they adopt specific behaviour which in turn leads to a spatial target.

4.2.2 Structure

As it has become clear in previous sections, the Spatially Transformed Social Networks model (STSN) is based on the power of ABMs. For the needs of this research, two slightly different versions of the model are built in the Processing 2.2.1.v. platform of Java computational language. The first one is designed to generate agents composing random graphs which could be depicted by a probability distribution or more simply, are generated by a random process. The second one is modified in a way that allows the import of data related to different types of social networks, like Balanced Tree (Ravasz, Barabási, 2003) and Barabasi-Albert (Barabási, Albert, 1999) models.

Diving deeper in the algorithm of the first version, there is a sequence of logical steps which is extended to the production of outputs. Analytically, at the beginning the model requires a population of nodes to be set and a maximum number of edges of which the new generated ER graph will consist. The model selects nodes randomly from the created dataset and link them two by two also in a random way. There is no rule that could restrict the number of potential edges per node. This process ensures the equal distribution of the probability for each vertex to be pair with all others. At the end of this phase, two lists are generated: a list with all nodes and a list with pairs of those nodes. The model spreads randomly the nodes on an empty plane. Nodes are called agents from this stage on.

In the next step, the model creates two more lists. The first one contains nodes which are characterised by a 2nd degree connection, that is, between the two nodes there is another node which is linked to both initial nodes. If the end-node of the pair p is the same with the start-node of the pair d and the end-node of the pair d is the start-point of the pair q and simultaneously the start-node of the pair d is not the end-node of the pair q, then the model generates pairs in which the start-node of the pair p and the end-node of the pair q are linked in three degrees depth. At the end of this process, the model has produced an ER network and has recorded all the potential pairs in three degrees of separation.

The preceding description depicts the construction of networks' structures and is considered as the first stage or input of the model. In the second version, the model instead of building nodes   from scratch and link them randomly, is modified to download and read data in CSV format. The source of the data and the preparation of them are described in a previous section (see 4.1.4.). After importing the dataset, this version follows exactly the same route with what was described earlier as the pair production process.

What follows is the establishment of the rules under which the agents are moving, acting and interacting with their environment as well as with their near or distant neighbours. These rules are quite simple but they produce quite complicated results.

To begin with, every agent is set to interact with the boarders of the plane and with each agent individually when the distance between them is lower than d. The length of d is a variable and is set to 25 pixels by default. Furthermore, the agents can interact in two ways depending on whether they are socially connected –in 1st, 2nd or 3rd degree of separation- or not. If they are not linked, agents just change direction of movement and in the next loop redefine their position and target. The first case is more complicated and is analysed below.

It is worth mentioning that agents do not move randomly. In each loop, every agent attempts to define a target towards which it moves. A target should be another agent with which they are pair, either in 1st, 2nd or 3rd degree. This behaviour derives from a social aspect called structural phenomenon and its branches Homophily and Propinquity which were examined in detail in the Literature Review chapter. In short, people with similar interests are more possible to be at similar locations and vice versa. In other words, linked agents attract each other. In a sense, this is a kind of gravity which applies forces to agents, who in turn make them moving. Although the amount of force is equal for every agent independently of the depth they are connected to, agents have preferences. For instance, if two agents are connected in the 1st degree, due to homophily they should be quite alike and thus due to propinquity ought to be at similar locations. However, the more the degree of social connectivity between two agents increases, 2nd or 3rd, the more the probability of agents to be spatially approached is decreased. In terms of modelling, this statement is interpreted as follows: the less the connectivity degree between agents is, the more able they are to detect each other. Thus, the STSN model introduces a variable called maximum distance of attractiveness which is 1000 pixels for 1st degree pairs, 100 pixels for 2nd degree pairs, 26 pixels for 3rd degree pairs by default. The numbers here should be considered as an attempt to show a trend.

So far, the way and the rules under which agents set their target, move and interact with their environment are analysed. What follows is the definition and description of two more important rules. The first one refers to the level of influence between two connected and spatially interacting agents. The second one gives shape to the final patterns.

As it was mentioned in a preceding section, the STSN model adopts the SNs' property of three degrees of influence, which was examined in detail in the literature review chapter. Briefly, the more the degree of social connectivity between two agents increases, 2nd or 3rd, the more the probability for an agent to be influenced from one another is decreased. In the STSN model, this property is expressed by the level of influence which is a variable. More specifically and as it is referred to in the literature review chapter, Christakis and Fowler (2009) found that the probability of an individual to be influenced by another with whom it is connected in the 1st degree is 52%. This probability for connections of 2nd degree is 25% and for 3rd degree relationships 15%. The STSN model adopts these percentages and sets them as the values of the level of influence variable by default. It is worth mentioning that these percentages could fluctuate depending on the case study of each research, that is, the location, the time, the social subject of the research and so on.

Finally, two are the possible operational situations for each agent. The first is already analysed and is the movement towards a moving target which is caused by homophily and propinquity phenomena via the maximum distance of attractiveness variable, and the second is standstill. When two agents are close enough and are socially connected, then the model generates a random number from 0 to 1 for each pair. If the agents constitute a 1st degree pair and the value of the generated number is less than 0.52, then the agents freeze and change their shape. Proportionally, the same applies to the other two types of relationships. The length of this particular distance between the two agents is determined by the minimum influence distance variable which is set to 40 pixels by default.

4.3 Building the spatial analysis

The purpose of this section is to introduce the reader to a quantitative way of measuring qualitative characteristics of the produced patterns. In general, the outputs of the STSN model could be characterised as abstract spatial forms or patterns. In this stage of research it cannot be assumed that these outputs represent a kind of real life settlements because the total process is quite conceptual. However, it is worth analysing spatial relationships among the produced abstract objects and interpret the results into quantitative statistics because this process bring to light unexplored measurable and comparable correlations between social and spatial structures. This is absolutely aligned with the aim of this research. Finally, by choosing appropriate spatial analysis platforms like DepthMapX software, quantitative spatial descriptions could lead to interesting conclusions like movement forecasting.

4.3.1 Quantifying the visibility of topology

This section contains information about how the abstract spatial patterns that emerged from the previous stage are used to export quantified measures. As architecture and urban design are considered as the fields in which the results of this research could be implemented, the spatial analysis tool should incorporate theoretical trends that are developed by and refer to these sectors.

The visual integration of space which is encapsulated in the Visibility Graph Analysis (VGA) method is a key concept for that purpose. The theory was developed in the context of space syntax by Alasdair Turner together with Alan Penn, David O'Sullivan and Maria Doxa (Turner et al., 2001a). According to this architectural theory, space, either as a building or an open urban space, could be understandable to users in visual terms and thus through visual relationships among built elements. The experience of space is directly linked with navigation in it and the latter depends more or less on visual steps. In other words, VGA assesses perceptual qualities of space and returns potential uses and movement patterns (Turner, 2004). Additionally, there are studies which correlate built environment visual relationships with standing still, discussing or simply occupying space (Doxa, 2001). Many of the qualitative characteristics of cities can be measured through VGA. For instance, Bill Hillier (2012) proposes that the sense of space integration is more efficient when the visual distance is long and the metric is short. In fact, the visual distance or visual integration of a point is considered as the number of visual steps that the system requires to get from that point to any other point within the system. The range of VGA seems to be wide in relation to spatial configurations and under this point of view, the correlation of VGA with SNA is crucial in order for the aim of this research to be satisfied.

Concluding, DepthMapX is considered as the most appropriate tool for performing VGA analyses, as it was originally programmed by Alasdair Turner in the context of his PhD, for that purpose. Although there is a debate on whether VGA actually fulfills its aims, the use of the visibility graph for the needs of this study is a promising path for more research. At least, VGA highlights spatial relationships between points and creates the background for understanding underlying processes related to spatial patterns.

What follows is the presentation of the key measurements of VGA which are used for the spatial analysis of the STSN model outputs. These topological measures are organised in two categories: global and local measurements. The main difference between them is related to neighbourhood size. Global measures indicate the visual relationship between a point and all the other points in the system or in other words, they prepare the shortest visual paths from each point to all other points while the second category focuses on visual relationships between a point and its neighbours.

4.3.2 Global measurements

First and foremost, a visibility graph is required to be built in order for a VGA to be performed. The information that a visibility graph contains and shares is the calculation of the connectivity of each point or simply, it calculates how many locations each point can see in the system. DepthMapX locates a grid over the map and calculates the values of the visibility for each cell which is called node.

I. Visual Entropy and Relativised Entropy

According to Turner (Tuner, 2001b), visual entropy and relativised entropy are measures which indicate the distribution of visual depths of nodes' locations rather than the metric depth. Entropy is low when the visual depth from a node is asymmetric and this is the case of locations visually close to a node. More evenly distributed visual depth increases the entropy. Relativised entropy is the expected visual distribution from a node. In real life experience, entropy indicates how easy it is for someone to traverse to a certain depth in a system. High entropy (low disorder system) means that it is easy while low entropy (high disorder) means it is hard.

II. Visual Integration

Visual Integration is an important measure in VGA and indicates how less visual steps is needed for a system in order to visual link all points to all others. When the visual distance is high the sense of space integration is more efficient (Hillier, 2012).

III. Visual Mean Depth

Mean depth is the sum of the shortest paths or the paths with the fewest changes in their direction in the visibility graph, which links each node to all the others within the graph, divided through by the number of nodes in the graph.

4.3.3 Local measurements

I. Visual Clustering Coefficient

The Visual Clustering Coefficient is related to junctions. Low clustering dictates where could be discovered a new area in the system. It is related to the existence of decision points (Turner, 2004).

II. Visual Control

III. Visual Controllability

4.3.4. Intelligibility

In space syntax, intelligibility is a property of spatial patterns which indicates how successful a navigation in unfamiliar urban environments could be (Hillier et al., 1987). Given the fact that people move purposefully in a city, their movements are in correlation with how good a picture of the entire urban form they have when they are in a local position. In a sense, intelligibility could be defined as how readable a pattern is by an agent and could be measured by correlating global and local visual properties of that pattern, e.g. global visual integration and connectivity (Conroy-Dalton, 2001). The r^2 of this relation reflects the intelligibility of the system and more specifically, the higher the r^2 is, the more readable, understandable and "user friendly" is the spatial pattern for a user.

4.4. Building the comparison between SNA and VGA

For the needs of the comparison between SNA and VGA statistics the produced datasets are visualised. The Numbers software used for building the bubble charts. This simple visualisation technique is considered the most meaningful and useful in order for the datasets to export information.

5. Findings

This chapter presents the outputs of the process described in detail in the previous chapter. As usual, first comes qualitative observations through the comparison of images which depict the structures of the particular social networks as well as the spatial patterns. The latter are observed under two filters. The first one is about their general geometry and some constant organisational principles they present in all the iterations during the calibration process. The second one is related to the VGA outputs. Subsequently, comparison charts are created in an attempt to bring to light specific quantitative relationships between these outputs and are organised in three categories. The first group of charts refers to quantitative characteristics of the particular social networks. The second group compares quantitative metrics of spatial characteristics which derive from the VGA outputs. Finally, in the third group of charts, there is an attempt to combine SNA and VGA outputs.

5.1 Qualitative Observations and Descriptions

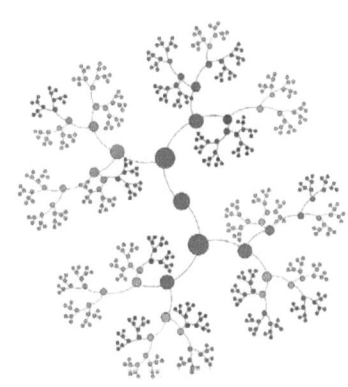

The statistics and the visualisation of the SNAs are exported from Gephi software, 0.8.2.beta, as it is mentioned in the Data Preparation section.

The first type of social network which is examined is the Hierarchical Network (HN) which is generated in Gephi. In the setup of the network there are two parameters: the degree of the root which was set to 2 and the

Figure 01. Hierarchical Network.

39

height of the tree which was set to 8. As it is shown in Figure 01, every node is connected to two other nodes and the last node is eight degrees separated from the central node. In the diagram there are clusters marked with different colours indicating the number and the position of local communities. The size of each node depicts the strength or the importance of each node and is correlated with its position in the system.

Figure 02. Barabasi-Albert network

The second type of social structure examined here, is the Barabasi-Albert (BA). Gephi asks the user to define the number of total nodes, which is set to 512 in this case, the number of nodes at the start time, set to 1, and the number of edges coming with every new node, which was set to 1 again. This means that the system generates one node at the beginning and adds a new one in every loop which in turn is connected with only one existed node. The algorithm gives a greater probability for the new node to be linked with a high degree existed node. The output is a social structure very close to real human networks. In Figure 02, there is a depiction of the network on which the STSN outcome is based. As in the HN model, the colours refer to different interlinked communities, the size of which depends on the position of the nodes and thus on the social structure of the system. In comparison to the HN model, there are more dense areas which compose sub-clusters. In other words, it seems that it consists of many autonomous sub-systems which are linked together through key-nodes, the bridges.

The last type of social networks that this book examines is the Erdos-Renyi (ER) model. This particular type of graph is generated by the STSN model in contrast to

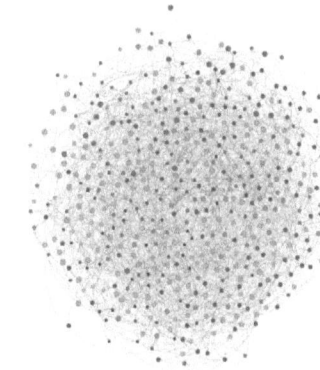

Figure 03. Erdos-Renyi network, strong relationships

the HN and BA graphs. The research introduces two versions of the ER network. Both have 512 nodes each, but the first has 2,200 edges and the second only 800 edges. The target is to highlight potential variations in terms of spatial patterns which could be caused by this difference. In a sense, there is an assumption that these two versions concern two types of social groups, a well-connected one with more social coherence - say Erdos-Renyi Strong (ERS)- and a less connected group with less social coherence -say Erdos-Renyi Weak (ERW). The null hypothesis is aligned with a common sense according to which the more interlinked the members of a group are, the denser and more clustered its spatial imprinting is expected to be.

Figure 04. Erdos-Renyi network, weak relationships

The visualisation and the statistics of both ER models are exported from Gephi. The STSN model, after running the simulation, exports a CSV file which contains the ID of each node and the edges the STSN model generated. The file in turn is imported to Gephi for visualisation and more analysis. In both sosiograms, there is no specific shape or an understandable and readable structure but instead, it seems that they look like abstract clouds. Figures 03 and 04 clearly depict that in the ERS graph the sub-communities are mixed in relation to the ERW graph in which the social communities are more clustered.

5.2 Quantitative Comparisons of social networks

What follows is a summary table and a series of bubble charts, the visualisation of which is designed to extract information about networks' specific attributes and mostly to bring to light comparative relationships among them. The importance of these charts is crucial because they allow the comparison of information which otherwise would stay compressed and fragmented in each network separately. Every chart combines values and refers to correlations between three points of data derived from table 01. The first point of data is always indicated in the X axis of the chart, the second point of data is always reflected in the Y axis of the chart and finally the third value, which is mentioned in the title of each chart, is correlated with the diameter of the bubble. The higher the value, the longest the diameter.

Table 01. SNA Summary Table

	Nodes	Edges	Av. Degree	Modularity	Communities	Density	Path Length	Nº of Shortest Paths	Diameter
Hierarchical Networks	511	510	1,996	0,908	24	0,004	12,117	260.610	16
Barabasi-Albert	512	511	1,996	0,892	24	0,004	6,333	261.632	17
Erdos-Renyi Strong Relationships	512	2.200	8,594	0,292	13	0,017	3,134	261.632	5
Erdos-Renyi Weak Relationships	512	800	3,125	0,588	36	0,006	5,388	236.690	12

Chart 01. Nodes, Edges & Average Degree

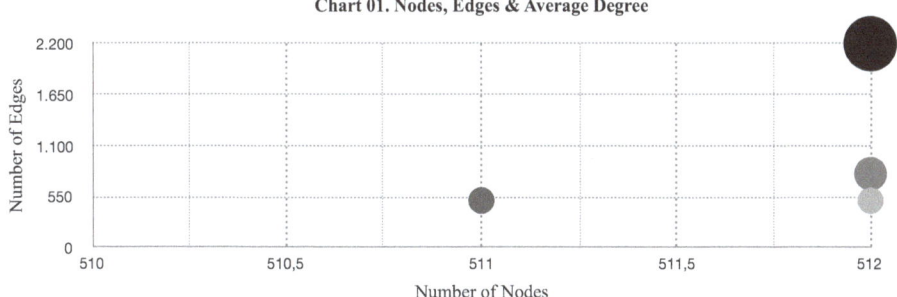

Hierarchical Network
Barabasi-Albert
Erdos-Renyi Strong Relationships
Erdos-Renyi Weak Relationships

Chart 01 is an introductory chart which presents the relationship between the number of nodes, the number of edges and the size of the average degree of all the types of social networks. It is shown that the ERS model has the higher average degree as well as more than twice the number of edges in relation to all other networks.

Chart 02. Modularity, Number of Communities & Density

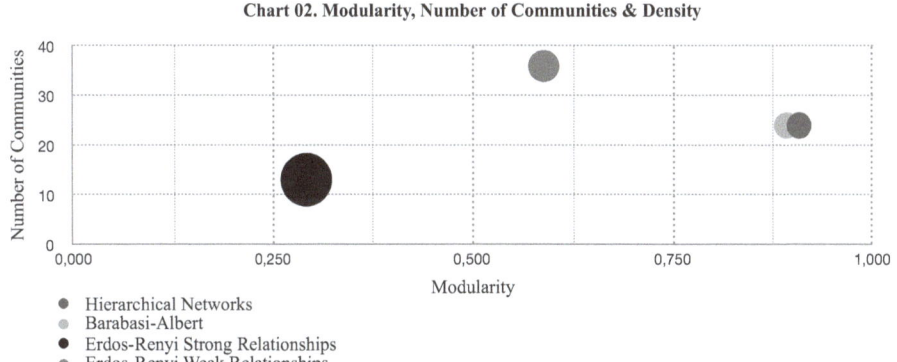

Hierarchical Networks
Barabasi-Albert
Erdos-Renyi Strong Relationships
Erdos-Renyi Weak Relationships

In Chart 02, the ERS random graph model, which in chart 01 seemed to have the higher number of edges, has the lower modularity among the others. That means that not only it has also the lower number of sub-communities and the higher density, but the connections between the nodes within its communities are not so dense. On the other hand, the connections between nodes of different sub-communities are stronger. In a sense, it seems to be a balanced network in terms of the spread of information. This is in contrast to what happens in the HN and BA models in which the strength of division of the networks into modules is high. Nodes within the modules create dense connections while the ties between the nodes which link the sub-communities are sparse. It seems that although they are two different structures, because of having the same density, they present similar behaviour in this sector. Finally, it is worth mentioning that regarding the ER random graphs of this study, their difference in density and in number of edges, seems to lead to an inverted relation in the modularity measurement. Generally, it is shown that more hierarchical structures lead to higher modularity.

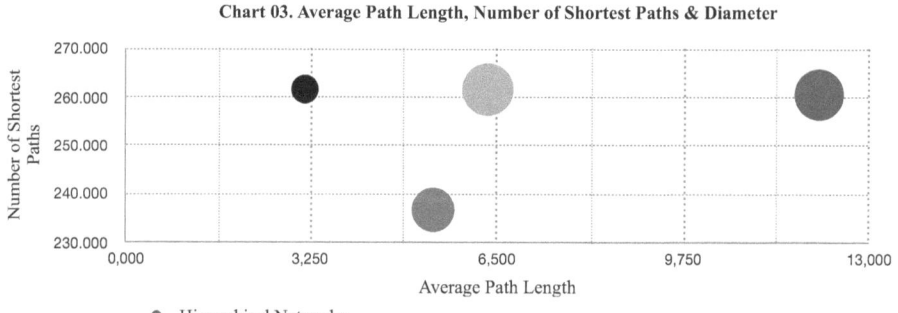

Chart 03. Average Path Length, Number of Shortest Paths & Diameter

The last chart (chart 03) shows that the most efficient structures of the social networks examined here, are the ER graphs because of their lower average path length. Although the population of the shortest paths are almost the same in the ERW, the BA and the HN models, the last one is much more inefficient than the others in terms of information flow. An interesting finding is also the fact that the ERW graph has equal population with the BA and HN models although it has almost 300 more edges which are linked randomly. In comparison with the previous chart (chart 03), higher density seems to mean lower diameter and vice versa. A difference between the last two diagrams which is worth noting, is that although the modularity and the number of sub-communities are almost the same in the cases of HN and BA models, the efficiency of a potential information flow on these networks appears to be unequal. Specifically, the BA structure seems to be almost two times more efficient than the HN graph. This is an important annotation given the fact that they have almost the same number of nodes and edges.

In summary, it seems that structure matters. HNs present the same dense connections between the nodes within modules with the BA networks but the difference in their architecture make the HNs more inefficient in the information flow. Additionally, the differences in the population of ties in the ER random graphs play both a significant role in their behaviour and also when compared to the behaviours of the rest types of networks.

What follows is the presentation of qualitative observations on the STSN model's outputs.

5.3 STSN model's outputs: Qualitative Observations

As it is mentioned in a previous chapter, STSN model uses as input data the topology of different types of social networks and after the process which was described in detail in the methodology chapter, exports spatial patterns, that is, patterns that essentially constitute the spatial transformations of the imported social networks.

Figure 05 depicts a typical environment of the STSN model simulation. The grey triangles in the left are the agents which are moving purposefully under specific rules. The shot is from the beginning of the simulation.

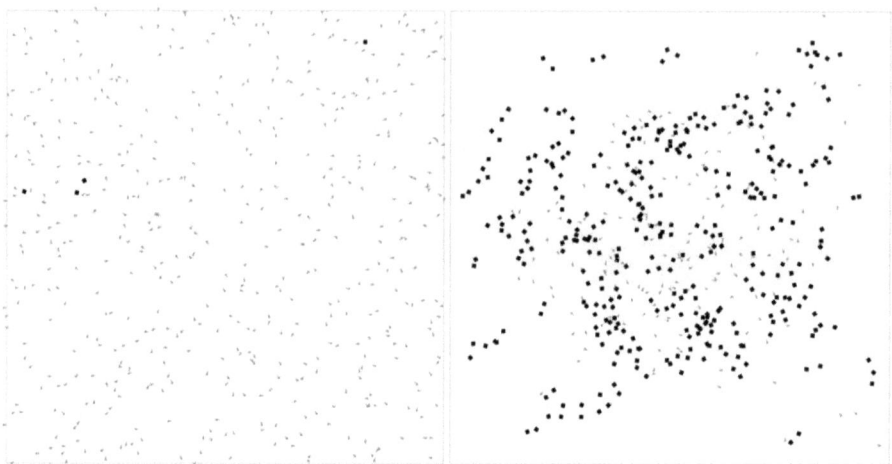

Figure 05. A typical environment of the STSN model simulation

The black squares are ex-agents which are steady and constitute the "built" environment. Each of the rest agents interact with the squares as well as with all the other agents until they get transformed to squares too. The following figures are the outputs which correspond to the already analysed social networks.

Figure 06. STHN

The first pattern (figure 06) is the spatial transformation of the HN graph (shortly, STHN). In a sense, the overview of the sketch seems to have been composed by abstract linear, but not straight, sub-structures from the centre to the periphery although the simulation in the STSN model started to build the diagram from the periphery to the centre. There are also big empty spaces between these sub-structures which are close to the centre and constitute a single set with the rest free space. In two or three places, small groups of higher density can be observed. Finally, it could be an abstract fractal synthesis similar to the outputs that the Diffusion Limited Aggregation model (DLA) produces (Witten et al., 1981). It is interesting to think this pattern in terms of real data. For example, in a sense, this could be a spatial allocation of military forces given the fact that the army has a strong administrative hierarchy.

Figure 07. STBA

The next figure (Figure 07) depicts the spatial transformation of the BA network (shortly, STBA). In contrast to the STHN pattern, the diffusion of the squares is more evenly spaced and occupies all the available area. In this case, there are also empty spaces which seem to be bigger in areas close to the periphery and smaller in areas close to the centre. Furthermore, it can be observed that there are some short lines composed by sequences of a small number of squares. In terms of real data, this pattern could reflect the spatial allocation of real social groups, given the fact that BA models are close to real social structures.

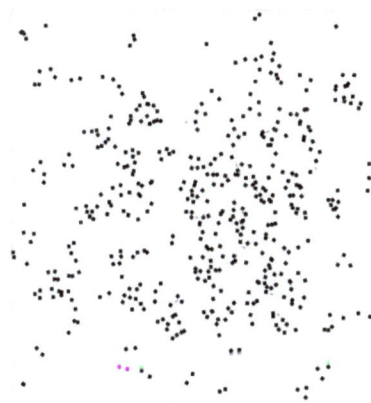

Figure 08. STERS

The third figure (Figure 08) shows the spatial transformation of the ERS random graph model (shortly, STERS). A very interesting observation is the fact that in a first sight there are some similarities with the STHN pattern. Indeed, there is a similar diffusion of the squares and it seems that there are enough empty spaces. But a closer look would show that it is more difficult for someone to recognise abstract linear sub-structures and clusters with higher density. Furthermore, the empty spaces do not shape the rest pattern as they do in the STHN pattern. Finally, the isolated dyads and triads in the periphery of the diagram are more in relation to figure 06. The STERS pattern could be the spatial allocation of a relatively coherent social group.

Figure 09. STERW

Figure 09 is the final one and depicts the spatial transformation of the ERW network (shortly, STERW). It seems that it is similar to the STHN pattern. The STERW consists of abstract linear, but not straight, sub-structures which seem to have been built from the centre to the periphery, although once again, the performance of the STSN model builds the form from the periphery to the centre. There is also a higher density in the centre of the pattern which consists of high density clusters as well as isolated dyads and triads of squares dispersed to the wider area. The difference from the STERS distribution is that STERW seems to be more dense and shrunken in total. This pattern could be the spatial allocation of a relatively incoherent social group. It is worth mentioning that in contrast to the null hypothesis (see 5.1. section), the more interlinked the members of a group are, the less dense and clustered its spatial transformation is expected to be, with the precondition that the members of the group are connected in a random way.

5.4 VGA outputs: Qualitative Observations

What is hoped to be presented here is a flavour of what the visibility graphs and the VGA of each social system could indicate. In this stage, the results are derived from qualitative observations on the VGA diagrams. Visibility graphs, as mentioned in a previous section, measure the visual connectivity of spatial patterns or simply calculate how many locations each point can see in the system. They are the base on which the VGA runs. These particular experiments were performed using the DepthMapX software. DepthMapX locates a grid over the spatial pattern and calculates the values of the visibility for each cell. The following diagrams use a grid of 5.000 metres as on one hand it is considered closer to the human scale and on the other, a grid of a smaller step would be extremely heavy for the computing power of a common computer. The DepthMapX colour spectrum starts from blue for low values and as the values arise, the colours become cyan, green, yellow, orange, red and magenta for high values.

In the appendix A there is a full set of similar diagrams which concern simulations based on global and local VGA measurements. The produced statistics of these simulations are examined and compared in the next section.

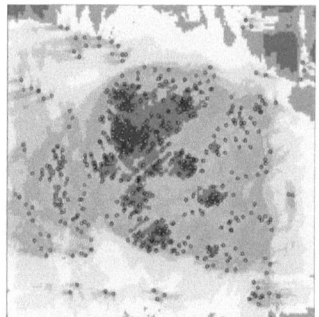

Diagram 01. Visibility graph of STHN

In this diagram there are more than one isolated, almost equal dark blue clusters of low connectivity in the centre of the image. That means that th8e mean length of spatial trips could be long. This phenomenon is expected since the majority of the objects are moved from the corners to the centre.

Diagram 02. Visibility graph of STBA

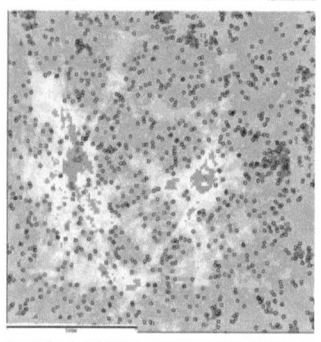

In contrast to the visibility graph of the STHN the diagram in left presents a totally inverted image. Here, there are two main and also many unequal in size clusters of high enough connectivity surrounded by lower and low visual connectivity areas. These areas exist also between the two main orange groups. It seems a more evenly distributed spatial transformation with relevant consequences in terms of mean trip length.

Diagram 03. Visibility graph of STERS

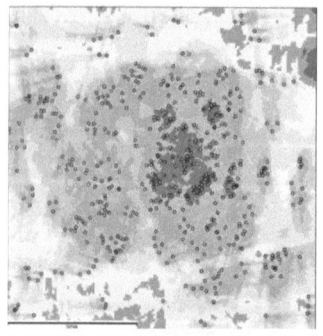

The predominant characteristic of this diagram is that the area with the lower connectivity is located in the centre. Although it seems similar to the STHN's visual graph, there is a softer fluctuation of connectivity values which is spatially expressed with wider transition zones.

Diagram 04. Visibility graph of STERW

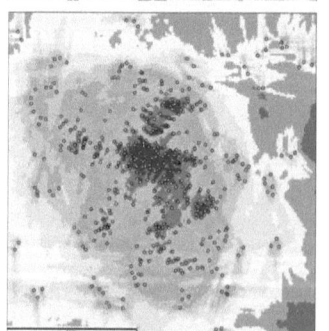

Less coherent social groups, the members of which are randomly connected, seems to have   spatial transformations with the most dense centres in terms of low visual connectivity. The transition zones from the central low value cluster to the high value corners are narrower than the respective of the STERS visibility graph.

5.5 Quantitative Comparisons of spatial patterns

Apart from the qualitative observations on the very interesting diagrams, DepthMapX exports datasets related to global and local measures of the spatially transformed social networks. The table 02 below presents the most significant of them. The following charts visualise this dataset and produce meaningful information about spatial characteristics among the spatial expressions of the social structures. In each chart, the X axis refers to the minimum value of the attribute, the Y axis refers to the maximum value of the attribute and the size of the bubble depicts the average value of that attribute.

Table 02. VGA Summary Table					
		Hierarchical Networks	Barabasi-Albert	Erdos-Renyi Strong Relationships	Erdos-Renyi Weak Relationships
Connectivity	Min.	35,000	3,000	15,000	10,000
	Max.	4.391,000	3.308,000	3.834,000	4.648,000
	Av. Value	2.367,840	1.475,040	2.168,910	2.735,360
Visual Integration	Min.	7,382	5,918	7,934	5,255
	Max.	18,243	14,114	17,908	19,595
	Av. Value	14,734	10,976	15,067	15,886
Visual Mean Depth	Min.	1,667	1,891	1,705	1,644
	Max.	2,649	3,125	2,590	3,402
	Av. Value	1,835	2,165	1,842	1,804
Visual Clustering Coefficient	Min.	0,185	0,190	0,201	0,187
	Max.	0,966	1,000	0,919	0,902
	Av. Value	0,499	0,326	0,434	0,513
Visual Control	Min.	0,039	0,005	0,013	0,023
	Max.	1,676	1,728	1,661	1,514
	Av. Value	1,000	1,000	1,000	1,000
Visual Controllability	Min.	0,005	0,003	0,003	0,004
	Max.	0,330	0,143	0,288	0,350
	Av. Value	0,181	0,070	0,164	0,208
Visual Entropy	Min.	0,160	0,316	0,160	0,153
	Max.	1,303	1,281	1,167	1,218
	Av. Value	0,752	0,959	0,676	0,786

Chart 04. Connectivity

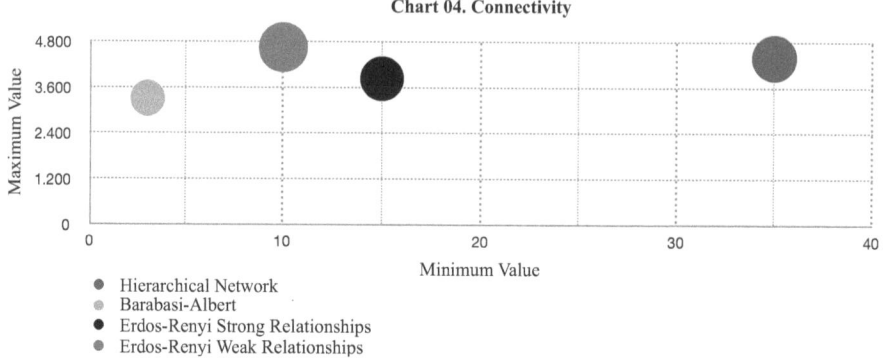

- Hierarchical Network
- Barabasi-Albert
- Erdos-Renyi Strong Relationships
- Erdos-Renyi Weak Relationships

The first chart (chart 04) shows the fluctuation of the connectivity value among the four spatial patterns. The extent to which visual connectivity indicates how many locations each point can see in the system, the STERW is the most visually connected and the less is the STBA pattern. While the minimum values are close to each other, the maximum values present greater fluctuation. This possibly is related to spatial density and dispersion which has in a sense opposite characteristics in the STERW and the STBA.

Chart 05. Visual Integration

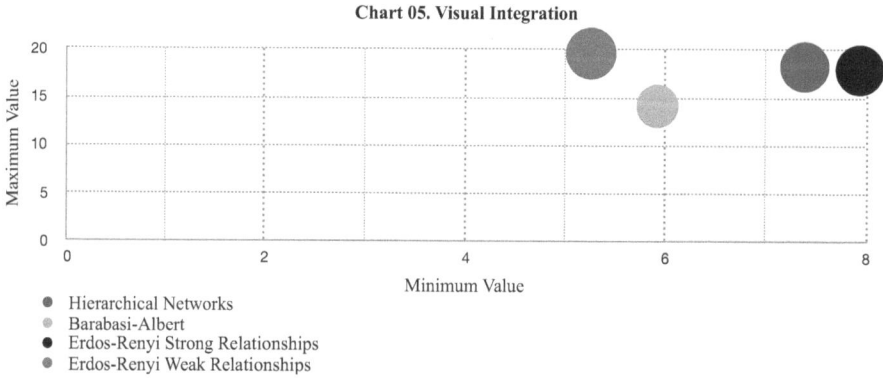

- Hierarchical Networks
- Barabasi-Albert
- Erdos-Renyi Strong Relationships
- Erdos-Renyi Weak Relationships

Visual integration or visual distance indicates how efficient the space integration is. High visual distances imply low metric distances. In this context, the STERW presents a more visual coherent spatial pattern in relation to the STBA (chart 05).

Again, this is aligned to the spatial density and dispersion of the objects in these patterns. The STHN and the STERS are almost equal in terms of visual integration.

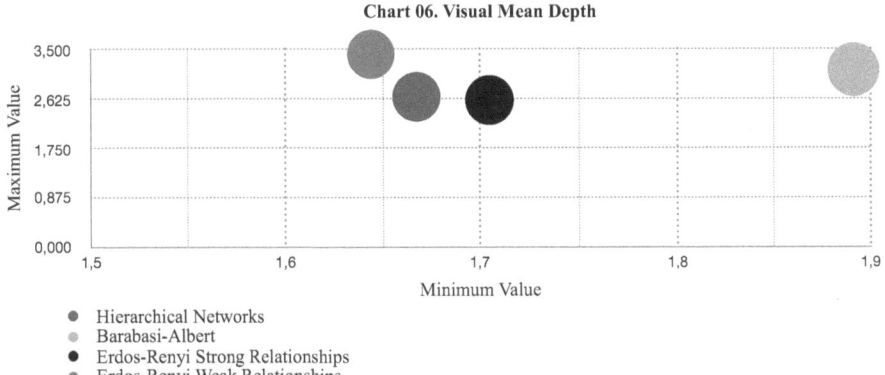

Visual mean depth dictates how visually connected a point is to all other points in the system. The extent to which high visual mean depth leads to low visually connected areas, the STBA seems to be the lower visually connected pattern in relation to the others which are rather equal (chart 06). The ERW has the highest maximum value and that means that it has some of the less visually connected areas of all the patterns.

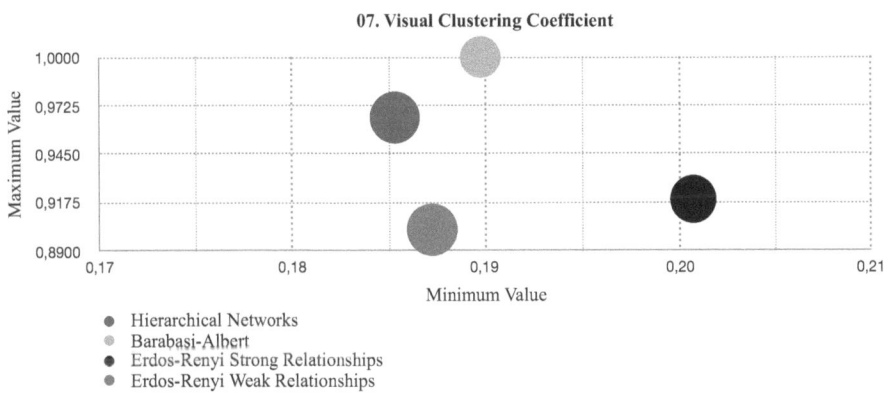

Visual clustering coefficient is a local measure related to the existence of junctions. It could perhaps dictate decision points and specifically, the lower the clustering coefficient is, the more decision points could be indicated. In this context, the STBA is the most complicated spatial pattern but due to its highest maximum value it seems that the potential decision points are not evenly distributed (chart 07). The STERW seems to have the shortest range of clustering coefficient values which could reflect high density and dispersion of the objects in the area.

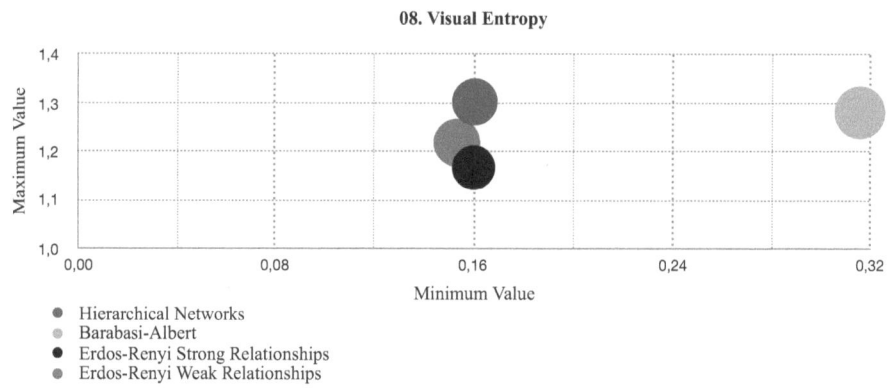

08. Visual Entropy

- Hierarchical Networks
- Barabasi-Albert
- Erdos-Renyi Strong Relationships
- Erdos-Renyi Weak Relationships

Visual entropy is a measure which indicates how ordered a system is and is related to the visual depth. In a sense, high entropy environments are these environments in which the visual depth is more evenly distributed and so it is easy to pass through. In this context, the STBA model seems to be the most accessible spatial pattern, followed by the STERW pattern. The STHN and the STERS are equal in terms of entropy but for the STHN pattern the range of values is greater. It seems that for the STHN case there are more areas in which the visual depth is more evenly distributed.

5.6 Intelligibility

As it is mentioned in a previous chapter, intelligibility could be measured by correlating global and local visual properties of a spatial pattern, e.g. global visual integration and connectivity. The higher the value of the R^2 of this correlation is, the more readable and well-navigated is the spatial pattern for a user even if the space is unfamiliar to him/her. The following scatterplots reflect the intelligibility of the four examined cases. The X axis refers to the visual connectivity of each spatial pattern and the Y axis is referred to the visual integration of them.

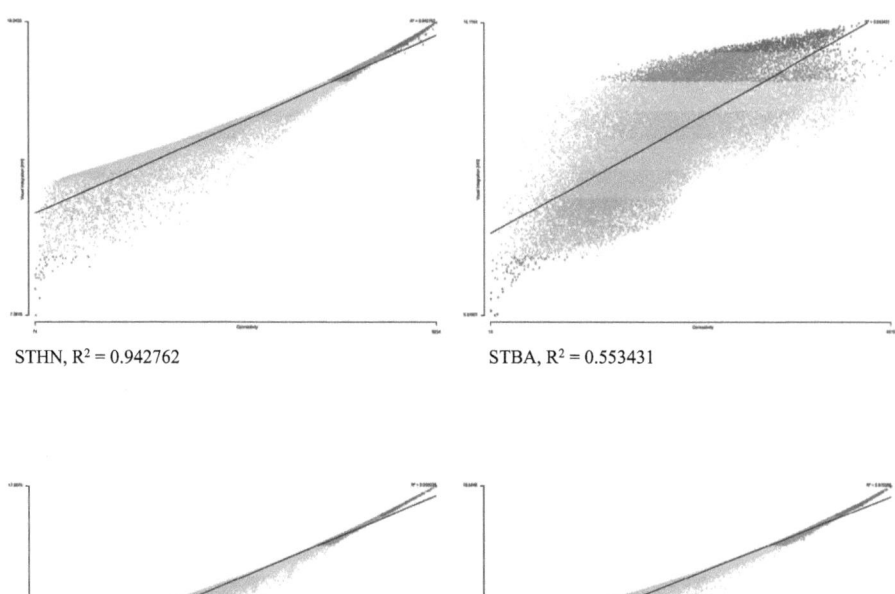

STHN, $R^2 = 0.942762$ STBA, $R^2 = 0.553431$

STERS, $R^2 = 0.956038$ STERW, $R^2 = 0.970068$

Under each scatterplot is the name of the particular spatially transformed social network and the value of the R^2.

5.7 Socio-Spatial Comparison

This section attempts to bring to light comparative relationships between social statistics and spatial measurements. It is believed that the produced information which derives from table 03 and the charts bellow, meets the target of this research and shapes a clearer image of its results. In the following charts, there are common variables in order for the comparison to be easier.

Table 03. SNA & VGA Comparative Table					
		Hierarchical Networks	Barabasi-Albert	Erdos-Renyi Strong Relationships	Erdos-Renyi Weak Relationships
Chart 09	Average Path Length	12,117	6,333	3,134	5,388
	Intelligibility	0,943	0,553	0,956	0,970
	Average Degree	1,996	1,996	8,594	3,125
Chart 10	Number of Communities	24,000	24,000	13,000	36,000
	Visual Clustering Coefficient	0,499	0,326	0,434	0,513
	Average Path Length	12,117	6,333	3,134	5,388
Chart 11	Diameter	16,000	17,000	5,000	12,000
	Visual Mean Depth	1,835	2,165	1,842	1,804
	Number of Communities	24,000	24,000	13,000	36,000

Chart 09 shows a comparison between social networks on the fields of the efficiency of the information flow, how readable is a spatial pattern from a user of the system and how well connected is a network as a whole.

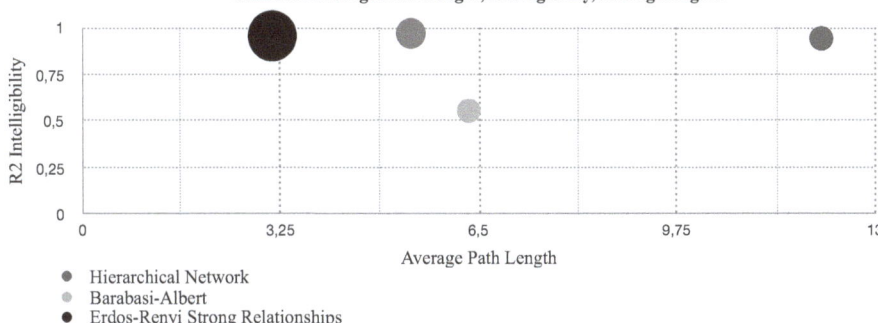

Chart 09. Average Path Length, Intelligibility, Average Degree

● Hierarchical Network
● Barabasi-Albert
● Erdos-Renyi Strong Relationships
● Erdos-Renyi Weak Relationships

To begin with, ERS, ERW and HN models present almost the same level of intelligibility and thus are equal in terms of space readability. In other words, the navigation of a user in a pattern could be successful even if he/she is unfamiliar with it. The HN graph, succeeds the readability of its spatial pattern although its members are weakly connected to all others and the efficiency of the information flow in the network is low. On the other hand, the fact that the BA network has the same average degree with the HN but almost the half of the HN's average path length, a difference which appears to be due to their different topology, seems to deprive from the BA model a high level of readability in its spatial transformation. Finally, differences in the average path length and average degree between the two versions of the ER random graphs, do not seem to influence how successful a navigation could be in their spatial transformations.

Chart 10 compares the earlier examined efficiency of information flow with the number of communities and the clustering coefficient of the spatially transformed social networks. The extent to which a number of communities imply low or high modularity, HN and BA networks consist of similar sub-systems in terms of density among their members and of how sparse are the links which connect the communities each other.

Chart 10. Number of Communities, Visual Clustering Coefficient, Average Path Length

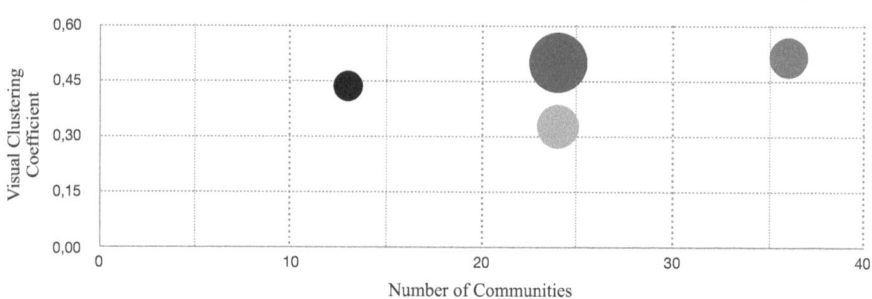

However, the STHN has higher visual clustering coefficient than the STBA and that dictates that perhaps STBA has a higher number of spatial decision points or sort of junctions. This notion bolsters the results of chart 09 according to which the STBA has lower level of space readability than the STHN. Additionally, although these two SNs have the same number of communities each, the efficiency of the information flow of the HN is lower. In summary, it seems that the denser the connections between members of communities are, the less decision points their spatial transformations have. Furthermore, the sparser the connections between communities are, the less the efficiency of the information flow in the system is. The same seems to apply to the ERS and ERW despite the fact that the ERS has a more than double number of nodes.

Chart 11 compares the number of social communities and the diameter of SNs with the visual mean depth of their spatial transformations or with how visually connected a point is to all other points of the produced pattern. The spatial pattern which is characterised by the higher value of visual mean depth and thus by lower visually connected space, is the STBA which is produced by the network with the longest diameter. This could perhaps dictate differences in navigation and movements of agents in the system.

Chart 11. Diameter, Visual Mean Depth, Number of Communities

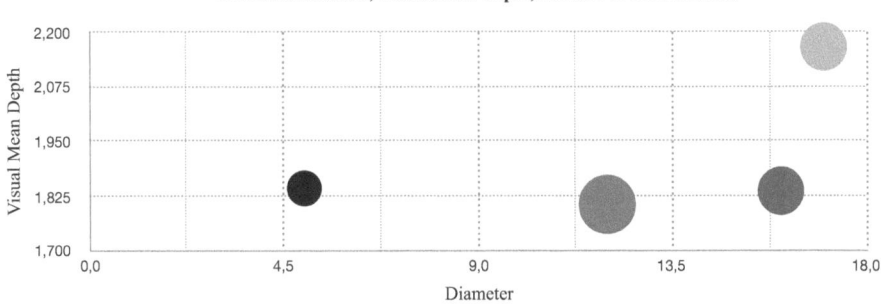

As it is mentioned in an earlier analysis, HN's communities are more strongly connected than those of BA's but they have sparser links with each other and this seems to have a relation with the notion above, although the number of modules are the same in the two networks. Slight differences in the visual mean depth between STHN, STERS and STERW do not follow big differences in the diameters of HN, ERS and ERW respectively.

6. Conclusions

6.1 Summary of the problem

The emergence of urban spatial patterns, in many cases, is subject to collective behaviours and when it is not, it is based on intuitive criteria of individuals. Although research in complexity theory of cities has become of age, only recently has the scientific community started to investigate the spatial dimension of social networks. However, it does not consider them as part of the urban design toolkit yet. The analysis of the results the STSN model produced, show that there is a promising research future in this direction as long as the articulation of social ties are an explanatory variable for the formation of space.

6.2 The main findings

First and foremost, the STSN model processes and produces a complex system since simple units with short-range relationships operate in parallel and generate emergent phenomena, complex behaviour and patterns. It is a non-linear procedure since change in initial condition e.g. a change in the percentages of influence, leads to a non-linear hang in outcome.

In the level of qualitative observations on the spatial patterns' topology, it was observed that there are some constant spatial characteristics for each social type, which were almost the same in every iteration of STSN's performance.

Very shortly, the STHN pattern is characterised by abstract linear, but not straight, sub-structures and big empty spaces between these which are close to the centre and constitute a single set with the rest free space. It is a kind of abstract fractal synthesis. The STBA pattern is characterised by a more evenly distributed spread of objects which occupy all the available area, by empty spaces which are bigger in areas close to the periphery and by some short lines composed by sequences of a small number of squares. The STERS pattern is similar to the STHN one but it is more difficult for someone to recognise abstract linear sub-structures and clusters with higher density. The empty spaces do not shape the rest pattern. The STERW pattern is characterised by higher density in the centre which consists of high density clusters as well as isolated dyads and triads of squares dispersed to the wider area. In comparison to all the other patterns, it is more dense and shrunken in total.

Furthermore, in an operational level, the quantitative research showed that the ERS, ERW and HN models produce spatial patterns which were characterised by a high level of readability or in other words they are in a sense more understandable in terms of navigation. Additionally, it seems that BA networks shape spatial patterns which are low visually connected and include many decision points or junctions, at least in relation to the other spatially transformed social networks.

Finally and in general, the more interlinked the members of a group are in random graphs, the less dense and clustered its spatial transformation is expected to be. Also, the denser the connections between the members in sub-communities are, the less decision points and thus the less complicated the spatial pattern is expected to be.

6.3 Implications of the research

Although any model is only as good as its assumptions, the results that the STSN model produced are encouraging. The findings are significant to the extent that they bring to light a promising research field which could lead to a new way for intervening in cities. For instance, if crime is correlated with high urban densities, instead of only increasing the security of these areas, a government could establish local social programmes that target to the enhancement of social connections. In military environments, where there are strong hierarchical social networks, the STSN model could optimise the distribution of ground forces by transforming the efficiency of the administrative structure into a spatial pattern. Furthermore, the extent to which citizens could shape the public space of their neighbourhoods, local urban environments could be redesigned by them in a sustainable way. By enhancing the relationships of local communities, local public space could decrease its functional embarrassed points. The model's ability to take into account and analyse real data, promises dynamic changes in the urban design sector. This is crucial for understanding what the role of social networks could be in the timeless process of urban development and for starting to think the city as a product of them.

6.4 Limitations

Given the fact that the proposed model opens a wide range research field, there are many limitations of this particular study as the depth of knowledge is limited and this, in turn, led to many assumptions and uncertainties. Under this point of view, the STSN model is so abstract that cannot reflect real case scenarios yet.

Another limitation of this particular study is related to computational power restrictions. Simulations could not be performed with high volume and power consuming data and so the model could not be calibrated for population more than 800 nodes.

6.5 Further research

A longer term research might be to identify whether these spatial forms share characteristics with real settlements. Additionally, more research should be done in order to optimise the variables of the model as well as to investigate more accurate values for the model's variables. Finally, it would be interesting if more research could be done regarding the percentages of influence between the degrees of separation for the case in which the nodes are not individuals but financial centres, political parties etc.

References

1. Adams, R.G. & Allan, G. (1999) Placing Friendship in Context. Cambridge ; New York, Cambridge University Press.

2. Albert, Réka; Barabási, Albert-László (2002). "Statistical mechanics of complex networks". Reviews of Modern Physics 74: 47–97. arXiv:cond-mat/0106096. Bibcode:2002RvMP...74...47A. doi:10.1103/RevModPhys.74.47

3. Al_Sayed, K., Turner, A., Hillier, B., Iida, S., (2014) (2nd Edition), "Space Syntax Methodology", Bartlett School of Graduate Studies, UCL, London.

4. Barabási, Albert-László; Albert, R. (1999) Emergence of scaling in random networks, Science 286 (5439): 509–512. arXiv:cond-mat/9910332. doi:10.1126/science.286.5439.509. PMID 10521342.

5. Bastian M., Heymann S., Jacomy M. (2009). Gephi: an open source software for exploring and manipulating networks. International AAAI Conference on Weblogs and Social Media.

6. Batty M., (2005) Cities and Complexity, Understanding Cities with Cellular Automata, Agent-based Models and Fractals, MIT, Cambridge, MA

7. Batty, M. (2008) Fifty years of urban modeling: macro-statics to micro-dynamics. In: The dynamics of complex urban systems. [Online]. Springer. pp. 1–20. Available from: http://link.springer.com/chapter/10.1007/978-3-7908-1937-3_1 [Accessed: 8 May 2014].

8. Batty, M. (2010) Generating cities from the bottom up. Embracing complexity in design. [Online] 1. Available from: http://books.google.com/books?hl=en&lr=&id=28WOAgAAQBAJ&oi=fnd&pg=PA1&dq=%22certain+goals+to+be+pursued.+We+conclude+with+various+demonstrations+of+how+idealised+plans+might+be%22+%22where+control+has+been+sufficiently+strict+to+enable+their+complete+implementation.+Most+development+in%22+&ots=nSAkCWw_pB&sig=LB-8QBeke_aSIl44zAkLOgKW0Pg [Accessed: 8 May 2014].

9. Bonabeau, E., (2002) Agent-based modelling: Methods and techniques for simulating human systems. Proceedings of the National Academy of Sciences, 99(90003), 7280-7287

10. Christakis, N. A. and Fowler, J. H. (2013), Social contagion theory: examining dynamic social networks and human behavior. Statist. Med., 32: 556–577. doi: 10.1002/sim.5408

11. Christakis, N., & Fowler, J. (2009). Connected: The Surprising Power of Our Social Networks and How They Shape Our Lives. New York: Back Bay Books; Reprint edition (January 12, 2011).

12. Cornoy Dalton, R., (2001) Spatial navigation in immersive virtual environments. PhD thesis, Bertlett School of Graduate Studies, University of London.

13. Diestel, Reinhard (2005), Graph Theory (3rd ed.), Berlin, New York: Springer-Verlag, ISBN 978-3-540-26183-4.

14. Dodds P.S. et al (2003) An experimental study of research in global social networks, Science 301: 827-829

15. Dorogovtsev, S.; Goltsev, A.; Mendes, J. (2002). Pseudofractal scale-free web. Physical Review E 65 (6). arXiv:cond-mat/0112143. Bibcode:2002PhRvE.. 65f6122D. doi:10.1103/PhysRevE.65.066122

16. Doxa M. (2001) Morphologies of co-presence and interaction in interior public space in places of performance: the Royal Festival Hall and the Royal National theatre, in Proceedings of the 3rd International Symposium on Space Syntax, Georgia Institute of Technology, Atlanta, Georgia, pp 27.1-27.13

17. Feld S., Carter W. (1998) Foci of activities as changing contexts for friendship. In Placing friendship in context, edited by R. G. Adams and G.Allan. Cambridge University, Cambridge

18. Festinger, L., Schacter S., Back K. (1950) Social pressures in informal groups: A study of human factors in housing. Stanford University Press, Stanford, Calif.

19. Franklin, S., Grasser, A., (1997) Is it an agent or just a program? A taxonomy for autonomous agents, In: Muller, J.P., Wooldridge, M.J., Jennings, N.R. (eds.)

Intelligent Agents III: Agent Theories, Architecture and Languages, Springer, Berlin, pp. 21-35.

20. Goleman, D. (2007) Social Intelligence: The New Science of Human Relationships. Reprint edition. New York, N.Y., Bantam.

21. Harary, Frank, Robert Z. Vorman, Dorwin Cartwright, (1965) Structural models: An introduction to the theory of directed graphs, Wiley, New York

22. Hillier, W. R. G. (2012). The genetic code for cities: is it simpler than we thought?. In J. Portugali, H. Meyer, E. Stolk, T. Ekim (Eds.), Complexity Theories of Cities Have Come of Age (pp. 129-152). Springer Verlag.

23. Hillier, B. & Hanson, J. (1984) The social logic of space. Cambridge [Cambridgeshire]; New York, Cambridge University Press.

24. Hillier, B., Burdett, R., Peponis, J., Penn, A., (1987) Creating life: or, does architecture determine anything? Arch. Behav. 3(3), 233-250

25. Kadushin, C. (2011) Understanding Social Networks: Theories, Concepts, and Findings. New York, OUP USA.

26. Kenrick, D.T., Goldstein, N.J. & Braver, S.L. (2012) Six Degrees of Social Influence: Science, Application, and the Psychology of Robert Cialdini. 1 edition. Oxford ; New York, OUP USA.

27. Kirke, D.M. (2009) Gender clustering in friendship networks: some sociological implications. Methodological Innovations Online. [Online] 4 (1), 23–36. Available from: http://www.pbs.plym.ac.uk/mi/pdf/17-04-09/3.%20kirk%20paper%2023-36.pdf [Accessed: 31 August 2014].

28. Kochen, M. (1989) The Small World: A Volume of Recent Research Commemorating Ithiel De Sola Pool, Stanley Milgram and Theodore Newcomb. Norwood, N.J, Ablex Publishing Corporation,U.S.

29. Korte, Bernhard; Lovász, László; Prömel, Hans Jürgen; Schrijver, Alexander (Eds.) (1990). Paths, Flows, and VLSI-Layout. Algorithms and Combinatorics 9, Springer-Verlag. ISBN 0-387-52685-4.

30. Lambiotte, R., Delvenne, J.-C. & Barahona, M. (2008) Laplacian dynamics and multiscale modular structure in networks. arXiv preprint arXiv:0812.1770.

[Online] Available from: http://arxiv.org/abs/0812.1770 [Accessed: 25 August 2014].

31. Lazarsfeld, Paul F., Merton Robert K. (1978) Freedom and control in modern society. In Friendship as a social process: A substantive and methodological analysis, edited by M. Berger, T. Abel and C. H. Page, Octagon Books, New York

32. Marshall, N., Batty, M. (2009) From darwinism to planning - Through Geddes and back. Town Country PLann 78(11), 462-464

33. McPherson M., Smith-Lovin L. and Cook M. (2001) Birds of a feather: Homophily in social networks. Annual Review of Sociology 27:415-44

34. Milgram S., Travers J. (1969) An experimental study in the small world problem, Sociometry 35, 4: 425-443

35. Moreno, Jacob lL. (1953) Who shall survive? Foundations of sociometry, group psychometry and sociodrama. Bacon House, New York. [Originally published as Nervous and Mental Disease Monograph 58, Washington, D. C., 1934]

36. Newman, Mark. E. J.; S. H. Strogatz and D. J. Watts (2001). "Random graphs with arbitrary degree distributions and their applications". Physical Review E 64 (026118). doi:10.1103/PhysRevE.64.026118

37. von Neumann, J., (1951) The general and logical theory of automata, in L.A. Jeffress, ed., Cerebral Mechanisms in Behavior – The Hixon Symposium, John Wiley & Sons, New York, pp. 1–31.

38. Pentland, A. (2014) Social Physics: How Good Ideas Spread—The Lessons from a New Science. New York, Penguin Press HC

39. Portugali, J. (2011) Complexity, cognition and the city / Juval Portugali ; with a foreword by Hermann Haken.

40. Portugali J., (2012), Complexity Theories of Cities: Achievements, Criticism and Potentials, DOI 10.1007/978-3-642-24544-2_4 Springer- Verlag Berlin Heidelberg.

41. Juval Portugali, Han Meyer, Egbert Stolk, & Ekim Tan (eds.) (2012) Complexity Theories of Cities Have Come of Age. [Online]. Berlin, Heidelberg, Springer Berlin Heidelberg. Available from: http://link.springer.com/ 10.1007/978-3-642-24544-2 [Accessed: 30 August 2014].

42.Ravasz, E. B.; Barabási, A. L. S. (2003). Hierarchical organization in complex networks, Physical Review E 67 (2). arXiv:cond-mat/0206130. Bibcode: 2003PhRvE..67b6112R. doi:10.1103/PhysRevE.67.026112

43.Smith, V. M. (2012) Social Networks Matter: A New Approach to Assess Clusters, Cultural Vitality, and their Implications for Planning Processes in Rio de Janeiro's Historic District, Praça Tiradentes. MS Thesis. Columbia University, New York.

44.Torrens, P. M. (2012) Moving Agent Pedestrians Through Space and Time. Annals of the Association of American Geographers, 102(1), 35-66

45.Turner A. (2004) Depthmap 4 - A researcher's handbook, Bartlett school of Graduate studies, UCL, London. http://www.vr.ucl.ac.uk/depthmap/handbook/depthmap4r1.pdf

46.Turner A., Doxa M., O'Sullivan D., Penn A. (2001a) From isovists to visibility graphs: a methodology for the analysis of architectural space. Environment and Planning B 28 (1): 103-121. doi: 10.1068/b2684

47.Turner A., (2001b) Depthmap: a program to perform visibility graph analysis, in Proceedings of the 3rd International Symposium on Space Syntax, Georgia Institute of Technology, Atlanta, Georgia

48.Verbrugge, Lois M., (1977) The structure of adult friendship choices. Social Forces 56:576-597

49.Verbrugge, Lois M.(1979). "Multiplexity in Adult Friendships", Social Forces, Vol. 57, No. 4 (Jun.), pp. 1286-1309

50.Wardhaugh, Ronald (2006). An Introduction to Sociolinguistics. New York: Wiley-Blackwell.

51.Witten T. A. , Sander L. M. (1981) Diffusion-Limited Aggregation, a Kinetic Critical Phenomenon, Phys. Rev. Lett. 47, 1400

52.Wong, L. H., Pattison, P., Robins, G., (2005) A spatial model for social networks. arXiv:physics/0505128v2

53.Xu, Guandong et al (2010). Web Mining and Social Networking: Techniques and Applications. Springer. p. 25. ISBN 978-1-4419-7734-2.

54.Zheng, X., Zhong, T., & Liu, M. (2009) Modelling crowd evaluation of a building based on seven methodological approaches. Building and Environment, 44(3), 437-445.

Appendix A

VGA diagrams for STHN / STBA / STERS / STERW
Visual Integration / Visual Entropy / Visual Integration [PValue] / Visual Integration
[Tekl] / Visual Mean depth / Visual Relativised Entropy

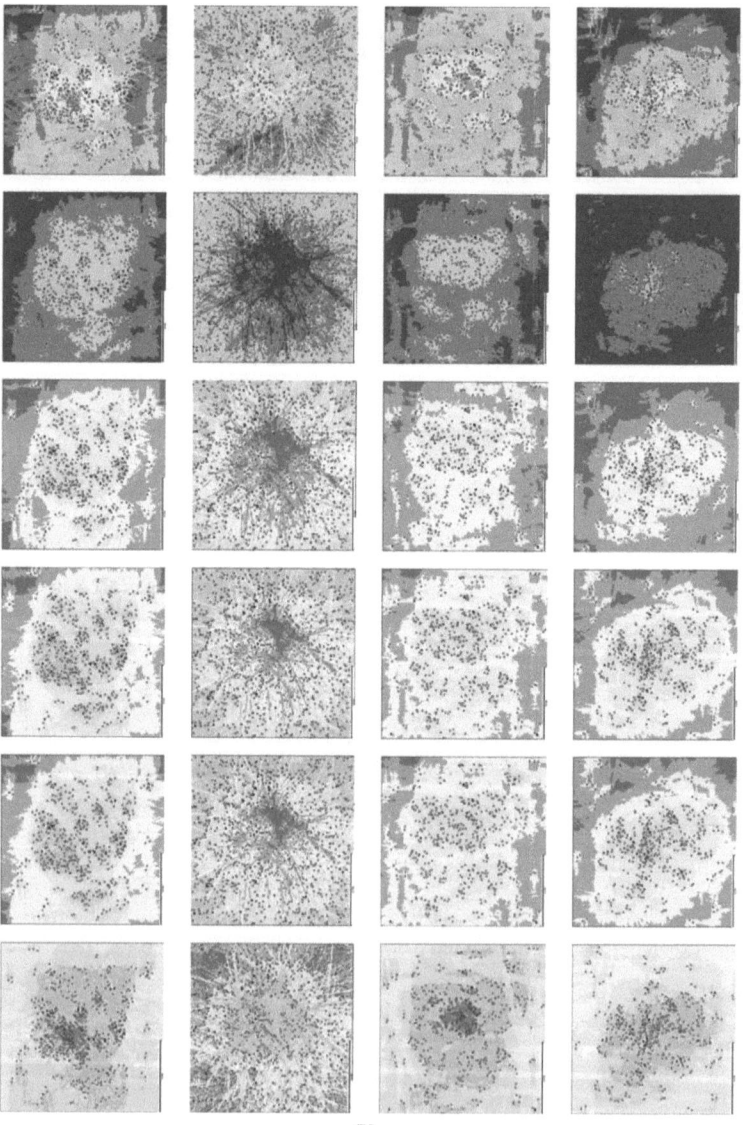

VGA diagrams for STHN / STBA / STERS / STERW

Visual Clustering Coefficient / Visual Control / Visual Controllability

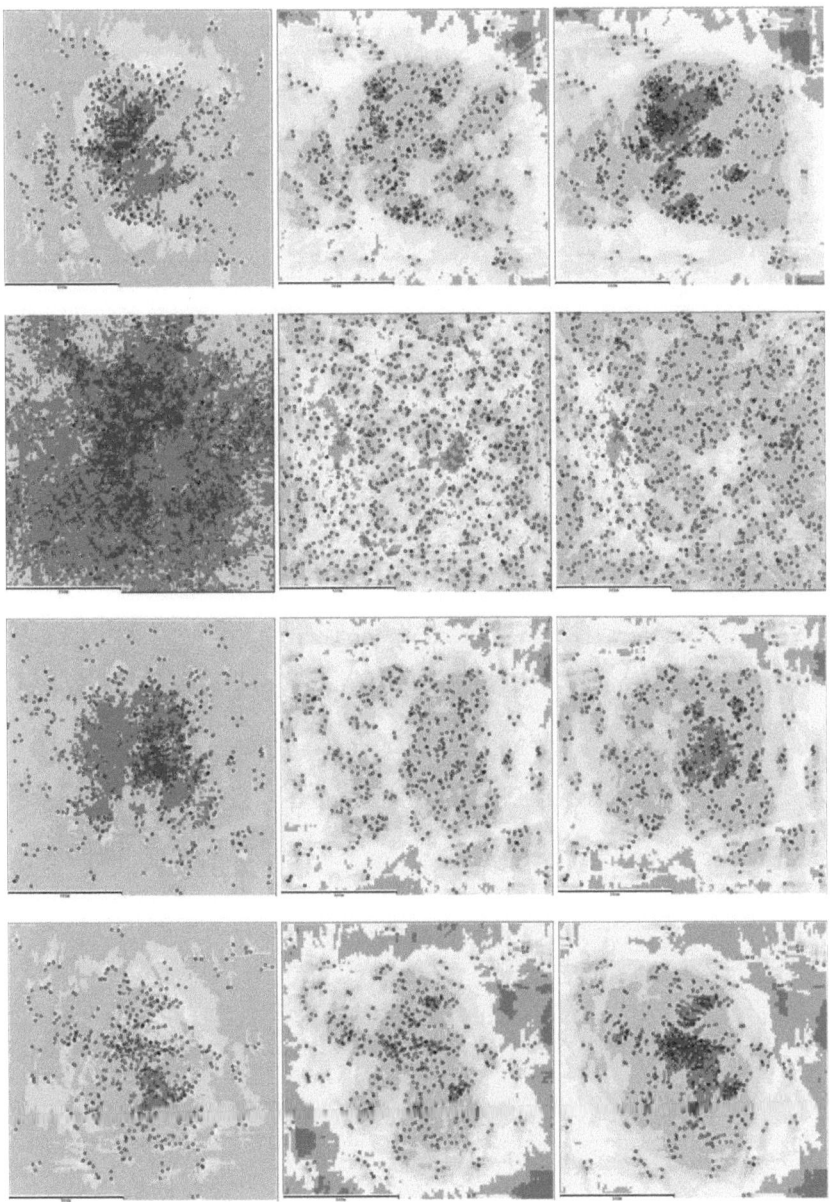

Appendix B

Diagramms for HN / BA / ERS / ERW

Degree Distribution / Betweenness Centrality / Closeness Centrality / Eccentricity

Centrality / Size Distribution

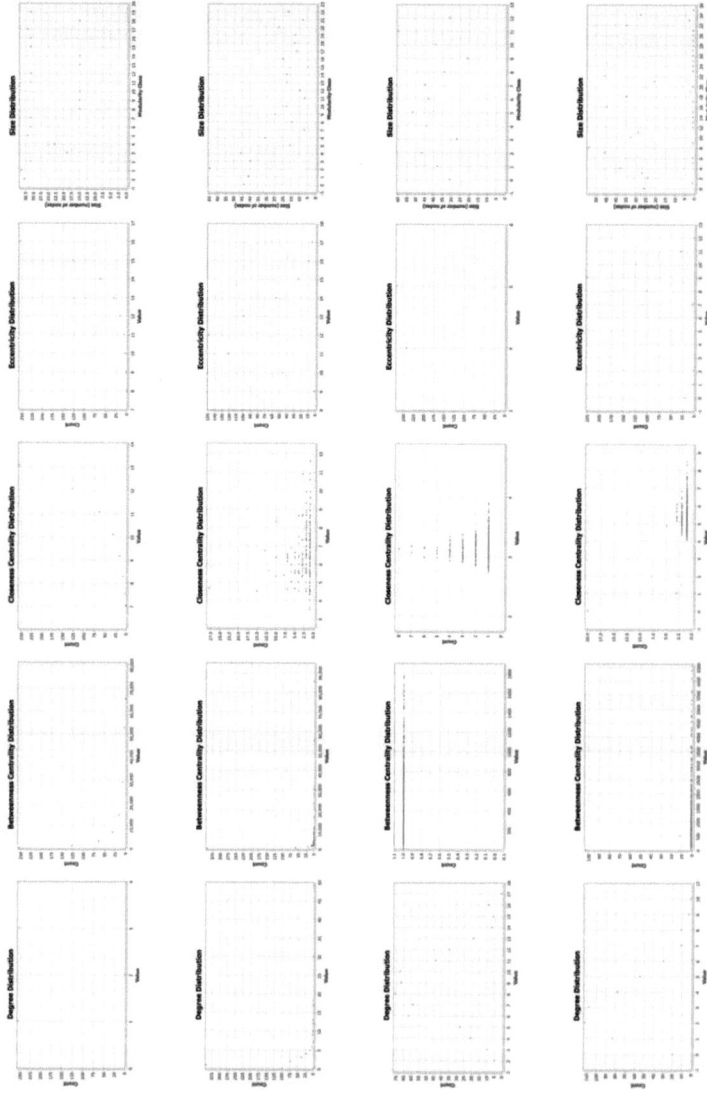

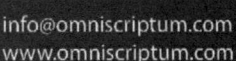

Printed by Books on Demand GmbH, Norderstedt / Germany